10-3-95

THE MYTH OF
THE FRAMEWORK

THE MYTH OF THE FRAMEWORK

In defence of science and rationality

Karl R. Popper

Edited by
M.A. Notturno

London and New York

First published 1994
by Routledge
11 New Fetter Lane, London EC4P 4EE

Simultaneously published in the USA and Canada
by Routledge
29 West 35th Street, New York, NY 10001

Typeset in 10/12pt Garamond by Solidus (Bristol) Limited
Printed and bound in Great Britain by
TJ Press Ltd, Padstow, Cornwall

British Library Cataloguing in Publication Data
A catalogue record for this book is available from the
British Library

Library of Congress Cataloging-in-Publication Data

Popper, Karl Raimund, Sir
The myth of the framework: in defence of science and
rationality/Karl Popper: edited by M.A. Notturno.
p.   cm.
Includes bibliographical references and index.
1. Science–Philosophy.   2. Physics–Philosophy.
3. Rationalism. 4. Reason
I. Notturno, Mark Amadeus.   II. Title.
Q175.P8633 1994
501–dc20   94–13562

ISBN 0–415–11320–2

To Werner Baumgartner

CONTENTS

AUTHOR'S NOTE, 1993

I do not regard myself as an expert in either science or philosophy. I have, however, tried hard all my life to understand something of the world we live in. Scientific knowledge, and the human rationality that produces it, are, I believe, always fallible, or subject to error. But they are, I believe, also the pride of mankind. For man is, so far as I know, the only thing in the universe that tries to understand what it is all about. May we continue to do so, and may we also be aware of the severe limitations of all our contributions.

For many years I have argued against intellectual fashions in the sciences, and even more against intellectual fashions in philosophy. The fashionable thinker is, in the main, a prisoner of his fashion, and I regard freedom, political freedom as well as a free and open mind, as one of the greatest, if not *the* greatest, value that our life can offer us.

Today it has become fashionable in the sciences to appeal to the specialized knowledge and authority of experts, and fashionable in philosophy to denigrate science and rationality. Oftentimes, this denigration of science and rationality is due to a mistaken *theory* of science and rationality – a theory which speaks of science and rationality in terms of specializations, experts, and authority. But science and rationality have really very little to do with special-ization and the appeal to expert authority. On the contrary, these intellectual fashions are actually an obstacle to both. For just as the fashionable thinker is a prisoner of his fashion, the expert is a prisoner of his specialization. And it is the freedom from intellectual fashions and specializations that makes science and rationality possible.

Today, the appeal to the authority of experts is sometimes excused by the immensity of our specialized knowledge. And it is sometimes defended by philosophical theories that speak of science and rationality in terms of specializations, experts, and authority. But in my view, the appeal to the authority of experts should be neither excused nor defended. It should, on the contrary, be recognized for what it is – an intellectual fashion – and it should be attacked by a frank acknowledgement of how little we know, and how much that little is due to people who have worked in many fields at the same time. And it should also be attacked by the recognition that the orthodoxy produced by intellectual fashions, specialization, and the appeal to authorities is the death of knowledge, and that the growth of knowledge depends entirely upon disagreement.

This is my excuse for collecting in a slim volume a few essays written in defence of science and rationality.

The essays in this volume were originally prepared on different occasions as lectures for non-specialist audiences. As a result, they often contained a résumé of my general approach to philosophy, and they sometimes contained brief discussions of some point taken up at greater length in another essay. This posed something of a problem when bringing them together in this volume. I have attempted to solve it by eliminating passages where there was an obvious overlap between two essays (provided this could be done without inflicting too much damage on their structure).

Some of what is said in these essays will undoubtedly be familiar to those who know my other works. But I believe that there is also much in these essays that will not be so familiar. And I have, in any event, tried hard to make every point and every argument as simple and as clear as I could.

This volume is permeated by a conviction which I have tried to indicate by its subtitle, and which has inspired my writings for at least the last sixty years. It is the conviction that scientific knowledge is, despite its fallibility, one of the greatest achievements of human rationality – and that we can, through the free use of our always fallible reason, nonetheless understand something about the world and, perhaps, even change it for the better.

K.R.P.
Kenley, Surrey, 1993

ACKNOWLEDGEMENTS

The idea of compiling this volume from my papers acquired in 1986 by the Hoover Institution at Stanford University, California, is due to my friend Dr Werner Baumgartner, a biochemist wishing to make my outlook known among American scientists. But he also took the initiative and kept it, carrying out his idea step by step. He obtained the financial help of the Ianus Foundation and engaged the scholarly help of Dr Mark Notturno, who in the later stages was assisted by his wife Kira. They enthusiastically worked on the task of selecting and editing these essays and lectures. I am deeply indebted to all who were concerned in the production of this volume.

K.R.P.
Kenley, Surrey, 17 March 1994

INTRODUCTION

All, or almost all, the papers collected in this volume are written to defend rationality and rational criticism. It is a way of thinking, and even a way of living: a readiness to listen to critical arguments, to search for one's own mistakes, and to learn from them. It is, fundamentally, an attitude that I have tried to formulate (perhaps first in 1932) in the following two lines:

'I may be wrong and you may be right,
and by an effort, we may get nearer to the truth.'

These two lines in italics here quoted were first printed in 1945 in my *Open Society* (volume II, the second page of chapter 24, 'The Revolt Against Reason'); and I italicized the lines in order to indicate that I regarded them as important. For these two lines were an attempt to summarize a very central part of my moral articles of faith. The view that they summed up I called 'critical rationalism'.

But the critics of my *Open Society* and of critical rationalism were, it seemed, blind to these two lines: so far as I know, none of my critics showed any interest in them, or quoted them. Some said that my book lacked any moral principle or ethical reasoning, others that my critical rationalism was dogmatic – too dogmatic; and there also was an attempt to replace my critical rationalism by a more radically critical and more explicitly defined position. But because this attempt bore the character of a definition, it led to endless philosophical arguments about its adequacy. I never found anyone who had taken notice of the two lines that I had intended as my moral *credo* – lines that had seemed to me to exclude any possibility of a dogmatic interpretation of 'critical rationalism'.

I am very ready to admit that all this is my own fault – that the two lines are obviously too brief to make the reader conscious of all that I shall attribute to them in the next paragraph: I hope you will agree that all I indicate there is indeed contained in these two lines – and more.

This is the reason why, after half a century, I am quoting them here. They were intended to contain, in a nutshell, a confession of faith, expressed simply, in unphilosophical, ordinary English; a faith in peace, in humanity, in tolerance, in modesty, in trying to learn from one's own mistakes; and in the possibilities of critical discussion. It was an appeal to reason; an appeal I had hoped would speak from all the pages of that long book.

It is perhaps of interest if I reveal that I owe the idea of formulating these two lines to a young Carinthian member of the National Socialist Party, neither a soldier nor a policeman, but wearing the Party uniform and a pistol. It must have been not long before the year 1933 – the year Hitler came to power in Germany – that this young man said to me: 'What, you want to argue? I don't argue: I shoot!' He may have planted the seed of my *Open Society*.

More than sixty years have gone by since this experience; and in the place where it happened, things seem to have improved. But over what was then the Carinthian-Yugoslavian border, a border that has not changed, the readiness to shoot under the pretext of ethnic provocation has increased in a terrible way. The attack of irrationalism upon argument has been going on through all these sixty years in more than sixty fashions. The pretext of ethnic provocation is the meanest and most revolting one of all, but not the newest. Quite possibly it is the oldest. This is not much comfort. But at least we need not accept that there is here – or elsewhere – a historical tendency for things to become worse. *The future depends upon ourselves*. It is we who bear all the responsibility.

For this reason, an important principle holds: *It is our duty to remain optimists*. Perhaps I should explain this in a few words before ending these notes.

The future is open. It is not predetermined and thus cannot be predicted – except by accident. The possibilities that lie in the future are infinite. When I say 'It is our duty to remain optimists', this includes not only the openness of the future but also that which all of us contribute to it by everything we do: we are all responsible for what the future holds in store.

Thus it is our duty, not to prophesy evil but, rather, to fight for a better world.

1

THE RATIONALITY
OF SCIENTIFIC
REVOLUTIONS
Selection *versus* instruction

The title of this series of Spencer lectures, 'Progress and Obstacles to Progress in the Sciences', was chosen by the organizers of the series. The title seems to imply that progress in science is a good thing, and that an obstacle to progress is a bad thing – a position held by almost everybody until quite recently. Perhaps I should make it clear at once that I accept this position, although with some slight and faintly obvious reservations to which I shall briefly allude later. Of course, obstacles which are due to the inherent difficulty of the problems tackled are welcome challengers. (Indeed, many scientists were greatly disappointed when it turned out that the problem of tapping nuclear energy was comparatively trivial, involving no new revolutionary change of theory.) But stagnation in science would be a curse. Still, I agree with Professor Bodmer's suggestion that scientific advance is only a *mixed* blessing[1]. Let's face it: blessings *are* mixed, though there may be some exceedingly rare exceptions.

My talk will be divided into two parts. The first part (sections I–VIII) is devoted to progress in science, and the second part (sections IX–XIV) to some of the social obstacles to progress.

Remembering Herbert Spencer, I shall discuss progress in science largely *from an evolutionary point of view* – more precisely, from the point of view of the theory of natural selection. Only the end of the first part (that is, section VIII) will be spent in discussing the

First published in *Problems of Scientific Revolution. Scientific Progress and Obstacles to Progress in the Sciences, The Herbert Spencer Lectures 1973*, edited by Rom Harré, Clarendon Press, Oxford, 1975. I wish to thank Troels Eggers Hansen, the Rev. Michael Sharratt, Dr Herbert Spengler, and Dr Martin Wenham for critical comments on this lecture.

progress of science *from a logical point of view*, and in proposing *two rational criteria* of progress in science, which will be needed in the second part of my talk.

In the second part I shall discuss a few obstacles to progress in science, more especially ideological obstacles. And I shall end (sections XI–XIV) by discussing the distinctions between, on the one hand, *scientific revolutions*, which are subject to rational criteria of progress, and, on the other hand, *ideological revolutions*, which are only rarely rationally defensible. It appeared to me that this distinction was sufficiently interesting to call my lecture 'The Rationality of Scientific Revolutions'. The emphasis here must be, of course, on the word 'scientific'.

I

I now turn to progress in science. I will be looking at progress in science from a biological or evolutionary point of view. I am far from suggesting that this is the most important point of view for examining progress in science. But the biological approach offers a convenient way of introducing the two leading ideas of the first half of my talk. They are the ideas of *instruction* and *selection*.

From a biological or evolutionary point of view, science, or progress in science, may be regarded as a means used by the human species to adapt itself to the environment: to invade new environmental niches, and even to invent new environmental niches.[2] This leads to the following problem.

We can distinguish between three levels of adaptation: genetic adaptation, adaptive behavioural learning, and scientific discovery (which is a special case of adaptive behavioural learning). My main problem in this part of my talk will be to enquire into the similarities and dissimilarities between the strategies of progress or adaptation on the *scientific* level and on those two other levels: *the genetic* level and the *behavioural* level. And I will compare the three levels of adaptation by investigating the role played on each level by *instruction* and by *selection*.

II

In order not to lead you blindfolded to the result of this comparison I will annnounce at once my main thesis. It is a thesis asserting the *fundamental similarity of the three levels*, as follows.

On all three levels – genetic adaptation, adaptive behaviour, and scientific discovery – the mechanism of adaptation is fundamentally the same.

This can be explained in some detail.

Adaptation starts from an inherited *structure* which is basic for all three levels: *the gene structure of the organism*. To it corresponds, on the behavioural level, the *innate repertoire* of the types of behaviour which are available to the organism and, on the scientific level, *the dominant scientific conjectures or theories*. These *structures* are always transmitted by *instruction*, on all three levels: by the replication of the coded genetic instruction on the genetic and the behavioural levels, and by social tradition and imitation on the behavioural and the scientific levels. On all three levels, the *instruction* comes from *within the structure*. If mutations, or variations, or errors occur, then these are new instructions, which also arise *from within the structure*, rather than *from without*, from the environment.

These inherited structures are exposed to certain pressures, or challenges, or problems: to selection pressures, to environmental challenges, to theoretical problems. In response, variations of the genetically or traditionally inherited *instructions* are produced[3] by methods which are at least partly *random*. On the genetic level, these are mutations and recombinations[4] of the coded instruction. On the behavioural level, they are tentative variations and recombinations within the repertoire. On the scientific level, they are new and revolutionary tentative theories. On all three levels we get new tentative trial instructions – or, briefly, tentative trials.

It is important that these tentative trials are changes that originate *within* the individual structure in a more or less random fashion – on all three levels. The view that they are *not* due to instruction from without, from the environment, is supported (if only weakly) by the fact that very similar organisms may sometimes respond in very different ways to the same new environmental challenge.

The next stage is that of *selecting* from the available mutations and variations: those of the new tentative trials which are badly adapted are eliminated. *This is the stage of the elimination of error.* Only the more or less well adapted trial instructions survive and are inherited in their turn. Thus we may speak of *adaptation by 'the method of trial and error'* – or better, by 'the method of trial and the elimination of error'. The elimination of error, or of badly adapted trial instructions, is also called '*natural selection*'. It is a kind of 'negative feedback' that operates on all three levels.

3

It is to be noted that in general *no equilibrium state of adaptation* is reached by any one application of the method of trial and the elimination of error, or by natural selection. First, because no perfect or optimal trial solutions to the problem are likely to be offered. Secondly – and this is more important – because the emergence of new structures, or of new instructions, involves a change in the environmental situation. New elements of the environment may become relevant. And in consequence, new pressures, new challenges and new problems may arise as a result of the structural changes which have arisen from within the organism.

On the genetic level the change may be a mutation of a gene, with a consequent change of an enzyme. Now the network of enzymes forms the more intimate environment of the gene structure. Accordingly, there will be a change in this intimate environment. And with it, new relationships between the organism and the more remote environment may arise – and further, new selection pressures.

The same happens on the behavioural level. For the adoption of a new kind of behaviour can be equated in most cases with the adoption of a new ecological niche. As a consequence, new selection pressures will arise, and new genetic changes.

On the scientific level, the tentative adoption of a new conjecture or theory may solve one or two problems. But it invariably opens up many *new* problems, for a new revolutionary theory functions exactly like a new and powerful sense organ. If the progress is significant then the new problems will differ from the old problems: the new problems will be on a radically different level of depth. This happened, for example, in relativity. It happened in quantum mechanics. And it is happening right now, most dramatically, in molecular biology. In each of these cases, new horizons of unexpected problems were opened up by the new theory.

This, I suggest, is the way in which science progresses. And our progress can best be gauged by comparing our old problems with our new ones. If the progress that has been made is great, then the new problems will be of a character undreamt-of before. There will be deeper problems, and there will be more of them. The further we progress in knowledge, the more clearly we can discern the vastness of our ignorance.[5]

I will now sum up my thesis.

On all the three levels which I am considering, the genetic, the behavioural, and the scientific levels, we are operating with inherited structures which are passed on by instruction – either through the genetic code or through tradition. On all the three levels, new structures and new instructions arise by trial changes from *within the structure*: by tentative trials, which are subject to natural selection or the elimination of error.

III

So far I have stressed the *similarities* in the working of the adaptive mechanism on the three levels. This raises an obvious problem: what about the *differences*?

The main difference between the genetic and the behavioural level is this. Mutations on the genetic level are not only random but completely 'blind' in two senses.[6] First, they are in no way goal directed. Secondly, the survival of a mutation cannot influence the further mutations, not even the frequencies or probabilities of their occurrence (though admittedly, the *survival* of a mutation may sometimes determine what kind of mutations may possibly *survive* in future cases). On the behavioural level, trials are also more or less random. But they are no longer completely 'blind' in either of the two senses mentioned. First, they are goal directed. And secondly, animals may learn from the outcome of a trial: they may learn to avoid the type of trial behaviour which has led to a failure. (They may even avoid it in cases in which it could have succeeded.) Similarly, they may also learn from success. And successful behaviour may be repeated, even in cases in which it is not adequate. However, a certain degree of 'blindness' is inherent in all trials.[7]

Behavioural adaptation is usually an intensely active process: the animal – especially the young animal at play – and even the plant are actively investigating the environment.[8]

This activity, which is largely genetically programmed, seems to me to mark an important difference between the genetic level and the behavioural level. I may here refer to the experience which the *Gestalt* psychologists call 'insight', an experience that accompanies many behavioural discoveries.[9] But it must not be overlooked that even a discovery accompanied by 'insight' may be *mistaken*: every trial, even one with 'insight', is of the nature of a conjecture or a hypothesis. Köhler's apes, it will be remembered, sometimes hit with 'insight' on what turns out to be a mistaken attempt to solve their problem. And even great mathematicians are sometimes misled by intuition. Thus animals and men have to try out their hypotheses. They have to use the method of trial and of error elimination.

On the other hand I agree with Köhler and Thorpe[10] that the trials of problem-solving animals are in general not completely blind. Only in extreme cases, when the problem which confronts the animal does not yield to the making of hypotheses, will the animal resort to more or less blind and random attempts in order to get out of a disconcerting situation. Yet even in these attempts, goal-directedness is usually discernible, in sharp contrast to the blind randomness of genetic mutations and recombinations.

Another difference between genetic change and adaptive behavioural change is that the former *always* establishes a rigid and almost invariable genetic structure. The latter, admittedly, leads *sometimes* also to a fairly rigid behaviour pattern which is dogmatically adhered to – radically so in the case of 'imprinting' (Konrad Lorenz) – but in other cases it leads to a flexible pattern which allows for differentiation or modification. For example, it may lead to exploratory behaviour, or to what Pavlov called the 'freedom reflex'.[11]

On the scientific level, discoveries are revolutionary and creative. Indeed, a certain creativity may be attributed to all levels, even to the genetic level: new trials, leading to new environments and thus to new selection pressures, create new and revolutionary results on all levels, even though there are strong conservative tendencies built into the various mechanisms of instruction.

Genetic adaptation can of course operate only within the time span of a few generations – at the very least, say, one or two generations. In organisms which replicate very quickly this may be a short time span, and there may be simply no room for behavioural adaptation. More slowly reproducing organisms are compelled to invent behavioural adaptation in order to adjust themselves to quick environmental changes. They thus need a behavioural repertoire, with types of behaviour of greater or lesser latitude or range. The repertoire, and the latitude of the available types of behaviour, can be assumed to be genetically programmed. And since, as indicated, a new type of behaviour may be said to involve the choice of a new environmental niche, new types of behaviour may indeed be genetically creative. For they may in their turn determine new selection pressures, and thereby indirectly decide upon the future evolution of the genetic structure.[12]

On the level of scientific discovery two new aspects emerge. The most important one is that scientific theories can be formulated linguistically, and that they can even be published. Thus they become objects outside of ourselves: objects open to investigation.

As a consequence, they are now open to *criticism*. Thus we can get rid of a badly fitting theory before the adoption of the theory makes us unfit to survive. *By criticizing our theories we can let our theories die in our stead*. This is of course immensely important.

The other aspect is also connected with language. It is one of the novelties of human language that it encourages story telling, and thus *creative imagination*. Scientific discovery is akin to explanatory story telling, to myth making and to poetic imagination. The growth of imagination enhances of course the need for some control, such as, in science, inter-personal criticism – the friendly hostile cooperation of scientists, which is partly based on competition and partly on the common aim to get nearer to the truth. This, and the role played by instruction and tradition, seems to me to exhaust the main sociological elements inherently involved in the progress of science – though more could be said of course about the social obstacles to progress, or the social dangers inherent in progress.

IV

I have suggested that progress in science, or scientific discovery, depends on *instruction* and *selection*: on a conservative or traditional or historical element, and on a revolutionary use of trial and the elimination of error by criticism, which includes severe empirical examinations or tests – that is, attempts to probe into the possible weaknesses of theories, attempts to refute them.

Of course, the individual scientist may wish to establish his theory rather than to refute it. But from the point of view of progress in science, this wish can easily mislead him. Moreover, if he does not himself examine his favourite theory critically, others will do so for him. The only results which will be regarded by them as supporting the theory will be the failures of interesting attempts to refute it – failures to find counter-examples where such counter-examples would be most expected in the light of the best of the competing theories. Thus it need not create a great obstacle to science if the individual scientist is biased in favour of a pet theory. Yet I think that Claude Bernard was very wise when he wrote: 'Those who have an excessive faith in their ideas are not well fitted to make discoveries.'[13]

All this is part of the critical approach to science, as opposed to the inductivist approach – of the Darwinian or eliminationist or selectionist approach, as opposed to the Lamarckian approach. The inductivist or Lamarckian approach operates with the idea of *instruction from without*, or from the environment. But the critical or Darwinian approach only allows *instruction from within* – from within the structure itself.

In fact, I contend that *there is no such thing as instruction from without the structure*, or the passive reception of a flow of information which impresses itself on our sense organs. All observations are theory-impregnated. There is no pure, disinterested, theory-free observation. (To see this, we may try, using a little imagination, to compare human observation with that of an ant or a spider.)

Francis Bacon was rightly worried about the fact that our theories may prejudice our observations. This led him to advise scientists that they should avoid prejudice by purifying their minds of all theories. Similar recipes are still given.[14] But to attain objectivity we cannot rely on the empty mind. Objectivity rests on criticism, on critical discussion, and on the critical examination of experiments.[15] And we must recognize, particularly, that our very sense organs incorporate what amount to prejudices. I have stressed before (in section II) that theories are like sense organs. Now I wish to stress that our sense organs are like theories. They *incorporate* adaptive theories (as has been shown in the case of rabbits and cats). And these theories are the result of natural selection.

V

However, not even Darwin or Wallace, not to mention Spencer, saw that there is no instruction from without. They did not operate with purely selectionist arguments. In fact, they frequently argued on Lamarckian lines.[16] In this they seem to have been mistaken. Yet it may be worth while to speculate about possible limits to Darwinism. For we should always be on the lookout for possible alternatives to any dominant theory.

I think that two points might be made here. The first is that the argument against the genetic inheritance of acquired characteristics (such as mutilations) depends upon the existence of a genetic mechanism in which there is a fairly sharp distinction between the gene structure and the remaining part of the organism: the soma. But this genetic mechanism must itself be a late product of evolution, and it was undoubtedly preceded by various other mechanisms of a less sophisticated kind. Moreover, certain very special kinds of mutilations *are* inherited – more particularly, mutilations of the gene structure by radiation. Thus if we assume that the primeval organism was a naked gene, then we can even say that every non-lethal mutilation to this organism would be inherited. What we cannot say is that this fact contributes in any way to an explanation of genetic adaptation, or of genetic learning – except indirectly, via natural selection.

The second point is this. We may consider the very tentative conjecture that, as a somatic response to certain environmental pressures, some chemical mutagen is produced, increasing what is called the spontaneous mutation rate. This would be a kind of semi-Lamarckian effect, even though *adaptation* would still proceed only by the elimination of mutations – that is, by natural selection. Of course, there may not be much in this conjecture, as it seems that the spontaneous mutation rate suffices for adaptive evolution.[17]

These two points are made here merely as a warning against too dogmatic an adherence to Darwinism. Of course, I do conjecture that Darwinism is right, even on the level of scientific discovery, and that it is right even beyond this level: that it is right even on the level of artistic creation. We do not discover new facts or new effects by copying them, or by inferring them inductively from observation, or by any other method of instruction by the environment. We use, rather, the method of trial and the elimination of error. As Ernst Gombrich says, 'making comes before matching':[18] the active production of a new trial structure comes before its exposure to eliminating tests.

VI

I suggest therefore that we conceive the way science progresses somewhat on the lines of Niels Jerne's and Sir Macfarlane Burnet's theories of antibody formation.[19] Earlier theories of antibody formation assumed that the antigen works as a negative template for the formation of the antibody. This would mean that there is *instruction from without*, from the invading antibody. The fundamental idea of Jerne was that the instruction or information which enables the antibody to recognize the antigen is, literally, inborn: that it is part of the gene structure, though possibly subject to a repertoire of mutational variations. It is conveyed by the genetic code, by the chromosomes of the specialized cells which produce the antibodies. And the immune reaction is a result of growth-stimulation given to these cells by the antibody-antigen complex.

Thus these cells are *selected* with the help of the invading environment (that is, with the help of the antigen), rather than instructed. (The analogy with the selection – and the modification – of scientific theories is clearly seen by Jerne, who in this connection refers to Kierkegaard, and to Socrates in the *Meno*.)

With this remark I conclude my discussion of the biological aspects of progress in science.

VII

Undismayed by Herbert Spencer's cosmological theories of evolution, I will now try to outline the cosmological significance of the contrast between *instruction from within the structure*, and *selection from without by the elimination of trials*.

To this end we may note first the presence, in the cell, of the gene structure, the coded instruction, and of various chemical sub-structures,[20] the latter in random Brownian motion. The process of instruction by which the gene replicates proceeds as follows. The various substructures are carried in random fashion (by Brownian motion) to the gene. Those which do not fit fail to attach themselves to the DNA structure. Those which fit, *do* attach themselves (with the help of enzymes). By this process of trial and selection,[21] a kind of photographic negative or complement of the genetic instruction is formed. Later, this complement separates from the original instruction and, by an analogous process, forms again its negative. This negative of the negative becomes an identical copy of the original positive instruction.[22]

The selective process underlying replication is a fast-working mechanism. It is essentially the same mechanism that operates in most instances of chemical synthesis, and also, especially, in processes like crystallization. Yet although the underlying mechanism is selective, and operates by random trials and by the elimination of error, it functions as a part of what is clearly a process of instruction rather than of selection. Admittedly, owing to the random character of the motions involved, the matching processes will be brought about each time in a slightly different manner. In spite of this, the results are precise and conservative: the results are essentially determined by the original structure.

If we now look for similar processes on a cosmic scale, a strange picture of the world emerges which opens up many problems. It is a dualistic world: a world of structures in chaotically distributed motion. The small structures (such as the so-called 'elementary particles') build up larger structures. And this is brought about mainly by chaotic or random motion of the small structures, under special conditions of pressure and temperature. The larger structures may be atoms, molecules, crystals, organisms, stars, solar systems, galaxies, and galactic clusters. Many of these structures appear to have a seeding effect, like drops of water in a cloud, or crystals in a solution. That is to say, they can grow and multiply by instruction. And they may persist, or disappear by selection. Some of them, such as the aperiodic DNA crystals[23] which constitute the gene structure of organisms and, with it, their building instructions, are almost infinitely rare and, we may perhaps say, very precious.

I find this dualism fascinating. I mean the strange dualistic picture of a physical world consisting of comparatively stable structures – or rather, structural processes – on all micro and macro levels, and of substructures on all levels in apparently chaotic or randomly distributed motion: a random motion that provides part of the mechanism by which these structures and substructures are sustained, and by which they may seed by way of instruction, and grow and multiply by way of selection and instruction. This fascinating dualistic picture is compatible with, yet totally different from, the well-known dualistic picture of the world as indeterministic in the small, owing to quantum-mechanical indeterminism, and deterministic in the large, owing to macro-physical determinism. In fact, it looks as if the existence of structures which do the instructing, and which introduce something like stability into the world, depends very largely upon quantum effects.[24] This seems to hold for structures on the atomic, molecular, crystal, organic, and even on the stellar levels (for the stability of the stars depends upon nuclear reactions), while for the supporting random movements we can appeal to classical Brownian motion and to the classical hypothesis of molecular chaos. Thus in this dualist picture of order supported by disorder, or of structure supported by randomness, the role played by quantum effects and by classical effects appears to be almost the opposite from that in the more traditional pictures.

VIII

So far I have considered progress in science mainly from a biological point of view. However, it seems to me that the following two logical points are crucial.

First, in order that a new theory should constitute a discovery or a step forward it should conflict with its predecessor – that is to say, it should lead to at least some conflicting results. But this means, from a logical point of view, that it should contradict[25] its predecessor: it should overthrow it.

In this sense, progress in science – or at least striking progress – is always revolutionary.

My second point is that progress in science, although revolutionary rather than merely cumulative,[26] is in a certain sense always conservative: a new theory, however revolutionary, must always be able to explain fully the success of its predecessor. In all those cases in which its predecessor was successful, it must yield results at least as good as those of its predecessor and, if possible, better results. Thus in these cases the predecessor theory must appear as a good approximation to the new theory, while there should be, preferably, other cases where the new theory yields different and better results than the old theory.[27]

The important point about the two logical criteria which I have stated is that they allow us to decide of any new theory, even before it has been tested, whether it will be better than the old one, provided it stands up to tests. But this means that, in the field of science, we have something like a criterion for judging the quality of a theory as compared with its predecessor, and therefore a criterion of progress. And so it means that progress in science can be assessed rationally.[28] This possibility explains why, in science, only progressive theories are regarded as interesting. And it thereby explains why, as a matter of historical fact, the history of science is by and large a history of progress. (Science seems to be the only field of human endeavour of which this can be said.)

As I have suggested before, scientific progress is revolutionary. Indeed, its motto could be that of Karl Marx: 'Revolution in permanence'. However, scientific revolutions are rational in the sense that, in principle, it is rationally decidable whether or not a new theory is better than its predecessor. Of course, this does not mean that we cannot blunder. There are many ways in which we can make mistakes.

An example of a most interesting mistake is reported by Dirac.[29] Schrödinger found, but did not publish, a relativistic equation of the electron, later called the Klein–Gordon equation, before he found and published the famous non-relativistic equation which is now called by his name. He did not publish the relativistic equation because it did not seem to agree with the experimental results as interpreted by the preceding theory. However, the discrepancy was due to a faulty interpretation of empirical results, and not to a fault in the relativistic equation. Had Schrödinger published it, the problem of the equivalence between his wave mechanics and the matrix mechanics of Heisenberg and Born might not have arisen, and the history of modern physics might have been very different.

It should be obvious that the objectivity and the rationality of progress in science is not due to the personal objectivity and rationality of the scientist.[30] Great science and great scientists, like great poets, are often inspired by non-rational intuitions. So are great mathematicians. As Poincaré and Hadamard have pointed out,[31] a mathematical proof may be discovered by unconscious trials, guided by an inspiration of a decidedly aesthetic character, rather than by rational thought. This is true, and important. But obviously, it does not make the result, the mathematical proof, irrational. In any case, a proposed proof must be able to stand up to critical discussion: to its examination by competing mathematicians. And this may well induce the mathematical inventor to check, rationally, the results which he reached unconsciously or intuitively. Similarly, Kepler's beautiful Pythagorean dreams of the harmony of the world system did not invalidate the objectivity, the testability, or the rationality of his three laws – nor the rationality of the problem which these laws posed for an explanatory theory.

With this, I conclude my two logical remarks on the progress of science. I will now move on to the second part of my lecture, and with it to remarks which may be described as partly sociological, and which bear on *obstacles* to progress in science.

IX

I think that the main obstacles to progress in science are of a social nature, and that they may be divided into two groups: economic obstacles and ideological obstacles.

On the economic side poverty may, trivially, be an obstacle (although great theoretical and experimental discoveries have been made in spite of poverty). In recent years, however, it has become fairly clear that affluence may also be an obstacle. Too many dollars may chase too few ideas. Admittedly, even under such adverse circumstances progress *can* be achieved. But the spirit of science is in danger. Big Science may destroy great science, and the publication explosion may kill ideas. Ideas, which are only too rare, may become submerged in the flood. The danger is very real, and it is hardly necessary to enlarge upon it, but I may perhaps quote Eugene Wigner, one of the early heroes of quantum mechanics, who sadly remarks:[32] 'The spirit of science has changed.'

This is indeed a sad chapter. But since it is all too obvious I shall not say more about the economic obstacles to progress in science. Instead, I will turn to discuss some of the ideological obstacles.

X

The most widely recognized of the ideological obstacles is ideological or religious intolerance, usually combined with dogmatism and lack of imagination. Historical examples are so well known that I need not dwell upon them. Yet it should be noted that even suppression may lead to progress. The martyrdom of Giordano Bruno and the trial of Galileo may have done more in the end for the progress of science than the Inquisition could do against it.

The strange case of Aristarchus and the original heliocentric theory opens perhaps a different problem. Because of his heliocentric theory Aristarchus was accused of impiety by Cleanthes, a Stoic. But this hardly explains the obliteration of the theory. Nor can it be said that the theory was too bold. We know that Aristarchus' theory was supported, a century after it was first expounded, by at least one highly respected astronomer (Seleucus).[33] And yet, for some obscure reason, only a few brief reports of the theory have survived. Here is a glaring case of the only too frequent failure to keep alternative ideas alive.

Whatever the details of the explanation, the failure was probably due to dogmatism and intolerance. But new ideas should be regarded as precious, and should be carefully nursed – especially if they seem to be a bit wild. I do not suggest that we should be eager to accept new ideas *just* for the sake of their newness. But we should be anxious not to suppress a new idea even if it does not appear to us to be very good.

There are many examples of neglected ideas, such as the idea of evolution before Darwin, or Mendel's theory. A great deal can be learned about obstacles to progress from the history of these neglected ideas. An interesting case is that of the Viennese physicist Arthur Haas who in 1910 partly anticipated Niels Bohr. Haas

published a theory of the hydrogen spectrum based on a quantization of J.J. Thomson's atom model: Rutherford's model did not yet
exist. Haas appears to have been the first to introduce Planck's
quantum of action into atomic theory with a view to deriving the
spectral constants. In spite of his use of Thomson's atom model,
Haas almost succeeded in his derivation. And as Max Jammer
explains in detail, it seems quite possible that the theory of Haas
(which was taken seriously by Sommerfeld) indirectly influenced
Niels Bohr.[34] In Vienna, however, the theory was rejected out of
hand. It was ridiculed and decried as a silly joke by Ernst Lecher
(whose early experiments had impressed Heinrich Hertz[35]), one of
the professors of physics at the University of Vienna, whose
somewhat pedestrian and not very inspiring lectures I attended some
eight or nine years later.

A far more surprising case, also described by Jammer,[36] is the
rejection in 1913 of Einstein's photon theory, first published in 1905,
for which he was to receive the Nobel prize in 1921. This rejection
of the photon theory formed a passage within a petition recom-
mending Einstein for membership of the Prussian Academy of
Science. The document, which was signed by Max Planck, Walther
Nernst, and two other famous physicists, was most laudatory, and
asked that a slip of Einstein's (such as they obviously believed his
photon theory to be) should not be held against him. This confident
manner of rejecting a theory which, in the same year, passed a severe
experimental test undertaken by Millikan, has no doubt a humorous
side. Yet it should be regarded as a glorious incident in the history
of science, showing that even a somewhat dogmatic rejection by the
greatest living experts can go hand in hand with a most liberal-
minded appreciation: these men did not dream of suppressing what
they believed was mistaken. Indeed, the wording of the apology for
Einstein's slip is most interesting and enlightening. The relevant
passage of the petition says of Einstein: 'That he may sometimes
have gone too far in his speculations, as for example in his
hypothesis of light quanta, should not weigh too heavily against
him. For nobody can introduce, even into the most exact of the
natural sciences, ideas which are really new, without sometimes
taking a risk.'[37] This is well said, but it is an understatement. One has
always to take the risk of being mistaken, and also the less important
risk of being misunderstood or misjudged.

However, this example shows, drastically, that even great scientists sometimes fail to reach that self-critical attitude which would

prevent them from feeling very sure of themselves while gravely misjudging things.

Yet a limited amount of dogmatism is necessary for progress. Without a serious struggle for survival in which the old theories are tenaciously defended, none of the competing theories can show their mettle – that is, their explanatory power and their truth content. Intolerant dogmatism, however, is one of the main obstacles to science. Indeed, we should not only keep alternative theories alive by discussing them, but we should systematically look for new alternatives. And we should be worried whenever there are no alternatives – whenever a dominant theory becomes too exclusive. The danger to progress in science is much increased if the theory in question obtains something like a monopoly.

XI

But there is an even greater danger: a theory, even a scientific theory, may become an intellectual fashion, a substitute for religion, an entrenched ideology. And with this I come to the main point of this second part of my lecture, the part that deals with obstacles to progress in science – to the distinction between scientific revolutions and ideological revolutions.

For in addition to the always important problem of dogmatism and the closely connected problem of ideological intolerance, there is a different and, I think, more interesting problem. I mean the problem which arises from certain links between science and ideology – links which do exist, but which have led some people to conflate science and ideology, and to muddle the distinction between scientific and ideological revolutions.

I think that this is quite a serious problem at a time when intellectuals, including scientists, are prone to fall for ideologies and intellectual fashions. This may well be due to the decline of religion, to the unsatisfied and unconscious religious needs of our fatherless society.[38] During my lifetime I have witnessed, quite apart from the various totalitarian movements, a considerable number of intellectually highbrow and avowedly non-religious movements with aspects whose religious character is unmistakable once your eyes are open to them.[39] The best of these many movements was that which was inspired by the father figure of Einstein. It was the best, because of Einstein's always modest and highly self-critical attitude and his humanity and tolerance. Nevertheless, I shall later have a few words

16

to say about what seem to me the less satisfactory aspects of the Einsteinian ideological revolution.

I am not an essentialist, and I shall not discuss here the essence or nature of 'ideologies'. I will merely state very generally and vaguely that I am going to use the term 'ideology' for *any non-scientific* theory, or creed, or view of the world which proves attractive, and which interests people, including scientists. (Thus there may be very helpful and also very destructive ideologies from, say, a humanitarian or a rationalist point of view.[40]) I need not say more about ideologies in order to justify the sharp distinction which I am going to make between science[41] and 'ideology', and further, between *scientific revolutions* and *ideological revolutions*. But I will elucidate this distinction with the help of a number of examples.

These examples will show, I hope, that it is important to distinguish between a scientific revolution in the sense of a rational overthrow of an established scientific theory by a new one, and all processes of 'social entrenchment' or perhaps 'social acceptance' of ideologies, including even those ideologies which incorporate some scientific results.

XII

As my first example I choose the Copernican and Darwinian revolutions, because in these two cases a scientific revolution gave rise to an ideological revolution. Even if we neglect here the ideology of 'Social Darwinism',[42] we can distinguish a scientific and an ideological component in both these revolutions.

The Copernican and Darwinian revolutions were *ideological* in so far as they both changed man's view of his place in the universe. They clearly were *scientific* in so far as each of them overthrew a dominant scientific theory: a dominant astronomical theory and a dominant biological theory.

It appears that the ideological impact of the Copernican and also of the Darwinian theory was so great because each of them clashed with a religious dogma. This was highly significant for the intellectual history of our civilization, and it had repercussions on the history of science (for example, because it led to a tension between religion and science). And yet, the historical and sociological fact that the theories of both Copernicus and Darwin clashed with religion is completely irrelevant for the rational evaluation of the scientific theories proposed by them. Logically it has nothing whatsoever to do with the *scientific* revolution started by each of the two theories.

It is therefore important to distinguish between scientific and ideological revolutions particularly in those cases in which the ideological revolutions interact with revolutions in science.

The example, more especially, of the Copernican ideological revolution may show that even an ideological revolution might well be described as 'rational'. However, while we have a logical criterion of progress in science – and thus of rationality – we do not seem to have anything like general criteria of progress or of rationality outside science (although this should not be taken to mean that outside science there are no such things as standards of rationality). Even a highbrow intellectual ideology which bases itself on accepted scientific results may be irrational, as is shown by the many movements of modernism in art (and in science), and also of archaism in art – movements which in my opinion are intellectually insipid since they appeal to values which have nothing to do with art (or science). Indeed, many movements of this kind are just fashions which should not be taken too seriously.[43]

Proceeding with my task of elucidating the distinction between scientific and ideological revolutions, I will now give several examples of major scientific revolutions which did not lead to any ideological revolution.

The revolution of Faraday and Maxwell was, from a scientific point of view, just as great as that of Copernicus, and possibly greater: it dethroned Newton's central dogma – the dogma of central forces. Yet it did *not* lead to an ideological revolution, though it inspired a whole generation of physicists.

J.J. Thomson's discovery (and theory) of the electron was also a major revolution. To overthrow the age-old theory of the indivisibility of the atom constituted a scientific revolution easily comparable to Copernicus' achievement: when Thomson announced it, physicists thought he was pulling their legs. But it did not create an ideological revolution. And yet, it overthrew both of the two rival theories which for 2400 years had been fighting for dominance in the field of matter – the theory of indivisible atoms, and that of the continuity of matter. To assess the revolutionary significance of this breakthrough it will be sufficient to remind you that it introduced structure as well as electricity into the atom, and thus into the constitution of matter. Also, the quantum mechanics of 1925 and 1926, of Heisenberg and of Born, of de Broglie, of Schrödinger and of Dirac, was essentially a quantization of the theory of the Thomson electron. And yet Thomson's scientific revolution did not lead to a new ideology.

Another striking example is Rutherford's overthrow in 1911 of the model of the atom proposed by J.J. Thomson in 1903. Rutherford had accepted Thomson's theory that the positive charge must be distributed over the whole space occupied by the atom. This may be seen from his reaction to the famous experiment of Geiger and Marsden. They found that when they shot alpha particles at a very thin sheet of gold foil, a few of the alpha particles – about one in twenty thousand – were reflected by the foil rather than merely deflected. Rutherford was incredulous. As he said later:[44] 'It was quite the most incredible event that has ever happened to me in my life. It was almost as incredible as if you fired a fifteen-inch shell at a piece of tissue paper and it came back and hit you.' This remark of Rutherford's shows the utterly revolutionary character of the discovery. Rutherford realized that the experiment refuted Thomson's model of the atom, and he replaced it by his nuclear model of the atom. This was the beginning of nuclear science. Rutherford's model became widely known, even among non-physicists. But it did not trigger off an ideological revolution.

One of the most fundamental scientific revolutions in the history of the theory of matter has never even been recognized as such. I mean the refutation of the electromagnetic theory of matter which had become dominant after Thomson's discovery of the electron. Quantum mechanics arose as part of this theory, and it was essentially this theory whose 'completeness' was defended by Bohr against Einstein in 1935, and again in 1949. Yet in 1934 Yukawa had outlined a new quantum-theoretical approach to nuclear forces which resulted in the overthrow of the electromagnetic theory of matter, after forty years of unquestioned dominance.[45]

There are many other major scientific revolutions which failed to trigger off any ideological revolution – for example, Mendel's revolution (which later saved Darwinism from extinction). Others are X-rays, radio-activity, the discovery of isotopes, and the discovery of superconductivity. To all these, there was no corresponding ideological revolution. Nor do I see as yet an ideological revolution resulting from the breakthrough of Crick and Watson.

XIII

Of great interest is the case of the so-called Einsteinian revolution – I mean Einstein's scientific revolution, which among intellectuals had an ideological influence comparable to that of the Copernican or Darwinian revolutions.

Of Einstein's many revolutionary discoveries in physics, there are two which are relevant here.

The first is special relativity, which overthrows Newtonian kinematics, replacing Galileo invariance by Lorentz invariance.[46] Of course, this revolution satisfies our criteria of rationality: the old theories are explained as approximately valid for velocities which are small compared with the velocity of light.

As to the ideological revolution linked with this scientific revolution, one element of it is due to Minkowski. We may state this element in Minkowski's own words. 'The views of space and time I wish to lay before you', Minkowski wrote, '... are radical. Henceforth space by itself, and time by itself, are doomed to fade away into mere shadows, and only a kind of union of the two will preserve an independent reality.'[47] This is an intellectually thrilling statement. But it is clearly not science: it is ideology. It became part of the ideology of the Einsteinian revolution. But Einstein himself was never quite happy about it. Two years before his death he wrote to Cornelius Lanczos: 'One knows so much and comprehends so little. The four-dimensionality with the [Minkowski signature of] +++− belongs to the latter category.'[48]

A more suspect element of the ideological Einsteinian revolution is the fashion of operationalism or positivism – a fashion which Einstein later rejected, although he himself was responsible for it, owing to what he had written about the operational definition of simultaneity. Although, as Einstein later realized,[49] operationalism is, logically, an untenable doctrine, it has been very influential ever since, in physics, and especially in behaviourist psychology.

With respect to the Lorentz transformations, it does not seem to have become part of the ideology that they limit the validity of the transitivity of simultaneity: the principle of transitivity remains valid within each inertial system while it becomes invalid for the transition from one system to another. Nor has it become part of the ideology that general relativity, or more especially Einstein's cosmology, allows the introduction of a preferred cosmic time and consequently of preferred local spatio-temporal frames.[50]

20

General relativity was, in my opinion, one of the greatest scientific revolutions ever, because it clashed with the greatest and best tested theory ever – Newton's theory of gravity and of the solar system. It contains, as it should, Newton's theory as an approximation, yet it contradicts it in several points. It yields different results for elliptic orbits of appreciable eccentricity. And it entails the astonishing result that any physical particle (photons included) which approaches the centre of a gravitational field with a velocity exceeding six-tenths of the velocity of light is not accelerated by the gravitational field, as in Newton's theory, but decelerated – that is, not attracted by a heavy body, but repelled.[51]

This most surprising and exciting result has stood up to tests, but it does not seem to have become part of the ideology.

It is this overthrow and correction of Newton's theory which, from a scientific (as opposed to an ideological) point of view, is perhaps most significant in Einstein's general theory. This implies, of course, that Einstein's theory can be compared point by point with Newton's[52] and that it preserves Newton's theory as an approximation. Nevertheless, Einstein never believed that his theory was true. He shocked Cornelius Lanczos in 1922 by saying that his theory was merely a passing stage: he called it 'ephemeral'.[53] And he said to Leopold Infeld[54] that the left-hand side of his field equation[55] (the curvature tensor) was as solid as a rock, while the right-hand side (the momentum–energy tensor) was as weak as straw.

In the case of general relativity, an idea which had considerable ideological influence seems to have been that of a curved four-dimensional space. This idea certainly plays a role in both the scientific and the ideological revolution. But this makes it even more important to distinguish the scientific from the ideological revolution.

However, the ideological elements of the Einsteinian revolution influenced scientists, and thereby the history of science. And this influence was not all to the good.

First of all, the myth that Einstein had reached his result by an essential use of epistemological, and especially operational, methods had, in my opinion, a devastating effect upon science. (It is irrelevant whether you get your results – especially good results – by dreaming them, or by drinking black coffee, or even from a mistaken epistemology.[56]) Secondly, it led to the belief that quantum mechanics, the second great revolutionary theory of the century, must outdo the Einsteinian revolution, especially with respect to its epistemological depth. It seems to me that this belief affected some of the great founders of quantum mechanics,[57] and also some of the great founders of molecular biology.[58] It led to the dominance of a subjectivist interpretation of quantum mechanics: an interpretation which I have been combating for almost forty years. I cannot here describe the situation, but while I am aware of the dazzling achievement of quantum mechanics (which must not blind us to the fact that it is seriously incomplete[59]) I suggest that the orthodox interpretation of quantum mechanics is not part of physics, but an ideology. In fact, it is part of a modernistic ideology, and it has become a scientific fashion which is a serious obstacle to the progress of science.

XIV

I hope that I have made clear the distinction between a scientific revolution and the ideological revolution which may sometimes be linked with it. The ideological revolution may serve rationality or it may undermine it. But it is often nothing but an intellectual fashion. Even if it is linked to a scientific revolution it may be of a highly irrational character, and it may consciously break with tradition.

But a scientific revolution, however radical, cannot really break with tradition, since it must preserve the success of its predecessors. This is why scientific revolutions are rational. By this I do not mean, of course, that the great scientists who make the revolution are, or ought to be, wholly rational beings. On the contrary: although I have been arguing here for the rationality of scientific revolutions, my guess is that should individual scientists ever become 'objective and rational' in the sense of 'impartial and detached', then we should indeed find the revolutionary progress of science barred by an impenetrable obstacle.

NOTES

1 Professor W.F. Bodmer concluded his Herbert Spencer Lecture, entitled 'Bio-medical Advances: A Mixed Blessing?', with the remark:

> Thus I believe that even if biomedical advances (and of course other scientific advances) are a mixed blessing – they are a blessing we cannot avoid and it is our job to see to it that the mixture comes out for the best. (See *Problems of Scientific Revolution. Scientific Progress*

1. THE RATIONALITY OF SCIENTIFIC REVOLUTIONS

and Obstacles to Progress in the Sciences, The Herbert Spencer Lectures 1973, p. 41.)

My own misgivings concerning scientific advance and stagnation arise mainly from the changed spirit of science, and from the unchecked growth of Big Science, which endangers great science. (See section IX of this lecture.) Biology seems to have escaped this danger so far, but not, of course, the closely related dangers of large-scale application.

2 The formation of membrane proteins, of the first viruses, and of cells, may perhaps have been among the earliest inventions of new environmental niches, though it is possible that other environmental niches (perhaps networks of enzymes invented by otherwise naked genes) may have been invented even earlier.

3 It is an open problem whether one can speak in these terms ('in response') about the genetic level (compare my conjecture about responding mutagens in section V). Yet if there were no variations, there could not be adaptation or evolution. And so we can say that the occurrence of mutations is either partly controlled by a need for them, or functions as if it were.

4 When in this lecture I speak, for brevity's sake, of 'mutation', the possibility of recombination is of course always tacitly included.

5 The realization of our ignorance has become pinpointed as a result, for example, of the astonishing revolution brought about by molecular biology.

6 For the use of the term 'blind' (especially in the second sense) see D.T. Campbell, 'Methodological Suggestions from a Comparative Psychology of Knowledge Processes', *Inquiry*, 2, 1959, pp. 152–82; 'Blind Variation and Selective Retention in Creative Thought as in Other Knowledge Processes', *Psychological Review*, 67, 1960, pp. 380–400; and 'Evolutionary Epistemology', in *The Philosophy of Karl Popper*, in The Library of Living Philosophers, edited by P.A. Schilpp, The Open Court Publishing Co., La Salle, Illinois, 1974, pp. 413–63.

7 While the 'blindness' of trials is relative to what we have found out in the past, randomness is relative to a set of elements (forming the 'sample space'). On the genetic level these 'elements' are the four nucleotide bases. On the behavioural level they are the constituents of the organism's repertoire of behaviour. These constituents may assume different weights with respect to different needs or goals, and the weights may change through experience (lowering the degree of 'blindness').

8 On the importance of active participation, see R. Held and A. Hein, 'Movement-Produced Stimulation in the Development of Visually Guided Behaviour', *Journal of Comparative Physiological Psychology*, 56, 1963, pp. 872–6. Cp. J.C. Eccles, *Facing Reality: Philosophical Adventures by a Brain Scientist*, Springer-Verlag, New York, 1970, pp. 66–7. The activity is, at least partly, one of producing hypotheses: see J. Krechevsky, '"Hypothesis" *versus* "Chance" in the Pre-Solution Period in Sensory Discrimination-learning', *University of California Publications in Psychology*, 6, 1932, pp. 27–44 (reprinted in *Animal*

Problem Solving, edited by A.J. Riopelle, Penguin Books, Harmondsworth, 1967, pp. 183–97).

9 I may perhaps mention here some of the differences between my views and the views of the *Gestalt* school of psychology. (Of course, I accept the fact of *Gestalt* perception. I am only dubious about what may be called *Gestalt* philosophy.)

I conjecture that the unity, or the articulation, of perception is more closely dependent on the motor control systems and the efferent neural systems of the brain than on afferent systems: that it is closely dependent on the behavioural repertoire of the organism. I conjecture that a spider or a mouse will never have insight (as had Köhler's ape) into the possible unity of the two sticks which can be joined together, because handling sticks of that size does not belong to their behavioural repertoire. All this may be interpreted as a kind of generalization of the James–Lange theory of emotions (1884; see William James, *The Principles of Psychology*, Macmillan & Co., London, 1890, volume II, pp. 449ff.), extending the theory from our emotions to our perceptions (especially to *Gestalt* perceptions) which thus would not be 'given' to us (as in *Gestalt* theory) but rather 'made' by us, by decoding (comparatively 'given') clues. The fact that the clues may mislead (optical illusions in man, dummy illusions in animals, etc.) can be explained by the biological need to impose our behavioural interpretations upon highly simplified clues. The conjecture that our decoding of what the senses tell us depends on our behavioural repertoire may explain part of the gulf that lies between animals and man. For through the evolution of human language our repertoire has become almost unlimited.

10 See W.H. Thorpe, *Learning and Instinct in Animals*, Methuen, London, 1956, pp. 99ff.; W. Köhler, *The Mentality of Apes*, Penguin, London, 1957, pp. 166ff.

11 See I.P. Pavlov, *Conditioned Reflexes*, Oxford University Press, London, 1927, especially pp. 11–12. In view of what he calls 'exploratory behaviour' and the closely related 'freedom behaviour' – both obviously genetically based – and of the significance of these for scientific activity, it seems to me that the behaviour of behaviourists who aim to supersede the value of freedom by what they call 'positive reinforcement' may be a symptom of an unconscious hostility to science. Incidentally, what B.F. Skinner (cp. his *Beyond Freedom and Dignity*, Alfred A. Knopf, New York, 1971) calls 'the literature of freedom' did not arise as a result of negative reinforcement, as he suggests. It arose, rather, with Aeschylus and Pindar, as a result of the victories of Marathon and Salamis.

12 Thus exploratory behaviour and problem solving create new conditions for the evolution of genetic systems, conditions which deeply affect the natural selection of these systems. One can say that once a certain latitude of behaviour has been attained – as it has been attained even by unicellular organisms (see especially the classic work of H.S. Jennings, *The Behavior of the Lower Organisms*, Columbia University Press, New York, 1906) – the initiative of the organism in selecting its ecology or habitat takes the lead, and natural selection within the new habitat

follows the lead. In this way, Darwinism can simulate Lamarckism, and even Bergson's 'creative evolution'. This has been recognized by strict Darwinists. For a brilliant presentation and survey of the history, see Sir Alister Hardy, *The Living Stream*, Collins, London, 1965, especially lectures VI, VII, and VIII, where many references to earlier literature will be found, from James Hutton (who died in 1797) onwards (see pp. 178f.). See also Ernst Mayr, *Animal Species and Evolution*, The Belknap Press, Cambridge, Mass., and Oxford University Press, London, 1963, pp. 604ff. and 611; Erwin Schrödinger, *Mind and Matter*, Cambridge University Press, Cambridge, 1958, chapter 2; F.W. Braestrup, 'The Evolutionary Significance of Learning', in *Videnskabelige Meddeleser Fra Dansk Naturhistorisk Forening*, *134*, 1971, pp. 89–102 (with a bibliography); and also my first Herbert Spencer Lecture, 1961, now in my *Objective Knowledge*.

13 Quoted by Jacques Hadamard, *The Psychology of Invention in the Mathematical Field*, Princeton University Press, Princeton, NJ, 1945, and Dover edition, New York, 1954, p. 48.

14 Behavioural psychologists who study 'experimenter bias' have found that some albino rats perform decidedly better than others if the experimenter is led to believe (wrongly) that the former belong to a strain selected for high intelligence. See Robert Rosenthal and Kermit L. Fode, 'The Effect of Experimenter Bias on the Performance of the Albino Rat', *Behavioral Science*, *8*, 1963, pp. 183–9. The lesson drawn by the authors of this paper is that experiments should be made by 'research assistants who do not know what outcome is desired' (p. 188). Like Bacon, these authors pin their hopes on the empty mind, forgetting that the expectations of the director of research may communicate themselves, without explicit disclosure, to his research assistants, just as they seem to have communicated themselves from each research assistant to his rats.

15 Cp. my *Logic of Scientific Discovery*, section 8, and my *Objective Knowledge*.

16 It is interesting that Charles Darwin in his later years believed in the occasional inheritance even of mutilations. See his *The Variation of Animals and Plants Under Domestication*, 2nd edition, John Murray, London, 1875, volume I, pp. 466–70.

17 Specific mutagens (acting selectively, perhaps on some particular sequence of codons rather than on others) are not known, I understand. Yet their existence would hardly be surprising in this field of surprises, and they might explain mutational 'hot spots'. At any rate, there seems to be a real difficulty in concluding from the absence of known specific mutagens that specific mutagens do not exist. Thus it seems to me that the problem suggested in the text (the possibility of a reaction to certain strains by the production of mutagens) is still open.

18 Cp. Ernst Gombrich, *Art and Illusion*, Pantheon Books, New York, 1960 and later editions; see the Index under 'making and matching'.

19 See Niels Kai Jerne, 'The Natural Selection Theory of Antibody Formation; Ten Years Later', in *Phage and the Origin of Molecular Biology*, edited by J. Cairns *et al.*, Cold Springs Harbor, New York,

1966, pp. 301–12; also 'The Natural Selection Theory of Antibody Formation', *Proceedings of the National Academy of Science, 41*, 1955, pp. 849–57; 'Immunological Speculations', *Scientific American, 229*, July 1973, pp. 52–60. See also Sir Macfarlane Burnet, 'A Modification of Jerne's Theory of Antibody Production, using the Concept of Clonal Selection', *Austrian Journal of Science, 20*, 1957, pp. 67–9; and *The Clonal Selection Theory of Acquired Immunity*, Cambridge University Press, Cambridge, 1959.

20 What I call 'structures' and 'substructures' are called 'integrons' by François Jacob, *The Logic of Living Systems: a History of Heredity*, Allen Lane, London, 1974, pp. 299–324.

21 Something might be said here about the close connection between 'the method of trial and the elimination of error' and 'selection'. All selection is error elimination. And what remains – after elimination – as 'selected' are merely those trials which have not been eliminated *so far*.

22 The main difference from a photographic reproduction process is that the DNA molecule is not two-dimensional but linear: a long string of four kinds of substructures ('bases'). These may be represented by dots coloured either red or green, or blue or yellow. The four basic colours are pairwise negatives (or complements) of each other. So the negative or complement of a string would consist of a string in which red is replaced by green, and blue by yellow – and *vice versa*. Here the colours represent the four letters (bases) which constitute the alphabet of the genetic code. Thus the complement of the original string contains a kind of translation of the original information into another yet closely related code. And the negative of this negative contains in turn the original information, stated in terms of the original (the genetic) code.

This situation is utilized in replication, when first one pair of complementary strings separates and when next two pairs are formed as each of the strings selectively attaches to itself a new complement. The result is the replication of the original structure, *by way of instruction*. A very similar method is utilized in the second of the two main functions of the gene (DNA) – the control, by way of instruction, of the synthesis of proteins. Though the underlying mechanism of this second process is more complicated than that of replication, it is similar in principle.

23 The term 'aperiodic crystal' (sometimes also 'aperiodic solid') is Schrödinger's. See his *What is Life?*, Cambridge University Press, Cambridge, 1944. Cp. his *What is Life?* in Schrödinger, *What is Life? & Mind and Matter*, Cambridge University Press, Cambridge, 1969, pp. 64 and 91.

24 That atomic and molecular structures have something to do with quantum theory is almost trivial, considering that the peculiarities of quantum mechanics (such as eigenstates and eigenvalues) were introduced into physics in order to explain the structural stability of atoms.

The idea that the structural 'wholeness' of biological systems has also something to do with quantum theory was first discussed, I suppose, in Schrödinger's small but great book *What is Life?* which, it may be said, anticipated both the rise of molecular biology and of Max Delbrück's influence on its development. In this book Schrödinger adopts a consciously ambivalent attitude towards the problem whether or not biology will turn out to be reducible to physics. In chapter 7, 'Is Life Based on the Laws of Physics?', he says (about living matter) first that 'we must be prepared to find it working in a manner that cannot be reduced to the ordinary laws of physics' (*What is Life? & Mind and Matter*, p. 81). But a little later he says that 'the new principle' (that is to say, 'order from order') 'is not alien to physics': it is 'nothing else than the principle of quantum physics again' (in the form of Nernst's principle) (*What is Life? & Mind and Matter*, p. 88). My attitude is also an ambivalent one. On the one hand, I do not believe in complete reducibility. On the other hand, I think that *reduction must be attempted*. For even though it is likely to be only partially successful, even a very partial success would be a very great success.

Thus my remarks in the text to which this note is appended (and which I have left substantially unchanged) were not meant as a statement of reductionism. All I wanted to say was that quantum theory seems to be involved in the phenomenon of 'structure from structure' or 'order from order'.

But my remarks were not clear enough. For in the discussion after the lecture Professor Hans Motz challenged what he believed to be my reductionism by referring to one of the papers of Eugene Wigner ('The Probability of the Existence of a Self-Reproducing Unit', chapter 15 of his *Symmetries and Reflections: Scientific Essays*, MIT Press, Cambridge, Mass., 1970, pp. 200–8). In this paper Wigner gives a kind of proof of the thesis that the probability is zero for a quantum theoretical system to contain a subsystem which reproduces itself. (Or, more precisely, the probability is zero for a system to change in such a manner that at one time it contains some subsystem and later a second subsystem which is a copy of the first.) I have been puzzled by this argument of Wigner's since its first publication in 1961. And in my reply to Motz I pointed out that Wigner's proof seemed to me refuted by the existence of Xerox machines (or by the growth of crystals) which must be regarded as quantum-mechanical rather than 'biotonic' systems. (It may be claimed that a Xerox copy or a crystal does not reproduce itself with sufficient precision. Yet the most puzzling thing about Wigner's paper is that he does not refer to degrees of precision, and that absolute exactness – which is not required – is, it seems, excluded at once by Pauli's principle.) I do not think that either the reducibility of biology to physics or else its irreducibility can be proved – at any rate not at present.

25 Thus Einstein's theory *contradicts* Newton's theory (although it contains Newton's theory as an approximation). In contradistinction to Newton's theory, Einstein's theory shows for example that in strong gravitational fields there cannot be a Keplerian elliptic orbit with appreciable eccentricity but without corresponding precession of the perihelion (as observed of Mercury).

26 Even the collecting of butterflies is *theory*-impregnated ('butterfly' is a *theoretical* term, as is 'water': it involves a set of expectations). The

recent accumulation of evidence concerning elementary particles can be interpreted as an accumulation of falsifications of the early electromagnetic theory of matter.

27 An even more radical demand may be made. For we may demand that if the apparent laws of nature should change, then the new theory, invented to explain the new laws, should be able to explain the state of affairs both before and after the change, and also the change itself, from universal laws and (changing) initial conditions. Cp. my *Logic of Scientific Discovery*, section 79, especially p. 253.

By stating these logical criteria for progress, I am implicitly rejecting the fashionable (anti-rationalistic) suggestion that two different theories such as Newton's and Einstein's are incommensurable. It may be true that two scientists with a verificationist attitude towards their favoured theories (Newtonian and Einsteinian physics, say) may fail to understand each other. But if their attitude is critical (as was Newton's and Einstein's) they will understand both theories, and see how they are related. See, for this problem, the excellent discussion of the comparability of Newton's and Einstein's theories by Troels Eggers Hansen in his paper 'Confrontation and Objectivity', *Danish Yearbook of Philosophy*, 7, 1972, pp. 13–72.

28 The logical demands discussed here (cp. chapter 10 of my *Conjectures and Refutations* and chapter 5 of *Objective Knowledge*), although they seem to me of fundamental importance, do not, of course, exhaust what can be said about the rational method of science. For example, in my *Postscript to the Logic of Scientific Discovery* I have developed a theory of what I call 'metaphysical research programmes' (see my *Realism and the Aim of Science*, edited by W.W. Bartley, III, Rowman & Littlefield, Totowa, New Jersey, 1983). This theory, it might be mentioned, in no way clashes with the theory of testing and of the revolutionary advance of science which I have outlined here. An example which I gave there of a metaphysical research programme is the use of the propensity theory of probability, which seems to have a wide range of applications.

What I say in the text should not be taken to mean that rationality depends on having a criterion of rationality. Compare my criticism of 'criterion philosophies' in Addendum I, 'Facts, Standards, and Truth', to volume II of my *Open Society*.

29 The story is reported by Paul A.M. Dirac, 'The Evolution of the Physicist's Picture of Nature', *Scientific American 208*, 1963, no. 5, pp. 45–53. See especially p. 47.

30 Cp. my criticism of the so-called 'sociology of knowledge' in chapter 23 of my *Open Society*, and pp. 155f. of my *Poverty of Historicism*.

31 Cp. Jacques Hadamard, *The Psychology of Invention in the Mathematical Field*, Princeton University Press, Princeton, NJ, 1945, and Dover edition, New York, 1954.

32 'A Conversation with Eugene Wigner', *Science*, *181*, 1973, pp. 527–33. See p. 533.

33 For Aristarchus and Seleucus see Sir Thomas Heath, *Aristarchus of Samos*, Clarendon Press, Oxford, 1913.

34 See Max Jammer, *The Conceptual Development of Quantum Mechanics*,

McGraw-Hill, New York, 1966, pp. 40–2.

35 See Heinrich Hertz, *Electric Waves*, Macmillan & Co., London, 1894; Dover edition, New York, 1962, pp. 12, 187f., 273.

36 See Jammer, *The Conceptual Development of Quantum Mechanics*, pp. 43f., and Théo Kahan, 'Un document historique de l'académie des sciences de Berlin sur l'activité scientifique d'Albert Einstein (1913)', *Archives internationales d'histoire des sciences*, *15*, 1962, pp. 337–42, especially p. 340.

37 Compare Jammer's slightly different translation in *The Conceptual Development of Quantum Mechanics*.

38 Our western societies do not, by their structure, satisfy the need for a father-figure. I discussed the problems that arise from this fact briefly in my (unpublished) William James Lectures in Harvard, 1950. My late friend, the psychoanalyst Paul Federn, showed me shortly afterwards an earlier paper of his devoted to this problem.

39 Fairly obvious examples are the roles of prophet played, in various movements, by Sigmund Freud, Arnold Schönberg, Karl Kraus, Ludwig Wittgenstein, and Herbert Marcuse.

40 There are many kinds of 'ideologies' in the wide and (deliberately) vague sense of the term I used in the text, and therefore many aspects to the distinction between science and ideology. Two may be mentioned here. One is that scientific theories can be distinguished or 'demarcated' (see note 41) from non-scientific theories which, nevertheless, may strongly influence scientists, and even inspire their work. (This influence, of course, may be good, or bad, or mixed.) A very different aspect is that of entrenchment: a scientific theory may function as an ideology if it becomes socially entrenched. This is why, when speaking of the distinction between scientific revolutions and ideological revolutions, I include among ideological revolutions changes in the social entrenchment of what may otherwise be a scientific theory.

41 In order not to repeat myself too often, I did not mention in this lecture my suggestion for a criterion of the empirical character of a theory (falsifiability or refutability as the criterion of demarcation between empirical theories and non-empirical theories). Since in English 'science' means 'empirical science', and since the matter is sufficiently fully discussed in my books, I have written things like the following (for example, in *Conjectures and Refutations*, p. 39): '... in order to be ranked as scientific, [statements] must be capable of conflicting with possible, or conceivable, observations'. Some people seized upon this like a shot (as early as 1932, I think). 'What about your own gospel?' is the typical move. (I found this objection again in a book published in 1973.) My answer to the objection, however, was published in 1934 (see my *Logic of Scientific Discovery*, chapter 2, section 10 and elsewhere). I may restate my answer: my gospel is not 'scientific', i.e., it does not belong to empirical science, but it is, rather, a (normative) *proposal*. My gospel (and also my answer) is, incidentally, criticizable, though not just by observation – and it has been criticized.

42 For a criticism of Social Darwinism, see my *Open Society*, chapter 10, note 71.

43 Further to my use of the vague term 'ideology' (which includes all kinds of theories, beliefs, and attitudes, including some that may influence scientists) it should be clear that I intend to cover by this term not only historicist fashions like 'modernism', but also serious, and rationally discussable, metaphysical and ethical ideas. I may perhaps refer to Jim Erikson, a former student of mine in Christchurch, New Zealand, who once said in a discussion: 'We do not suggest that science invented intellectual honesty, but we do suggest that intellectual honesty invented science.' A very similar idea is to be found in chapter 9 of Jacques Monod's book *Chance and Necessity*, Alfred A. Knopf, New York, 1971. See also my *Open Society*, chapter 24. We might say, of course, that an ideology which has learned from the critical approach of the sciences is likely to be more rational than one which clashes with science.

44 Lord Rutherford, 'The Development of the Theory of Atomic Structure', in *Background of Modern Science*, edited by J. Needham and W. Pagel, Cambridge University Press, Cambridge, 1938, pp. 61–74. The quotation is from p. 68.

45 See my 'Quantum Mechanics without "The Observer"', in *Quantum Theory and Reality*, edited by Mario Bunge, Springer-Verlag, New York, 1967, esp. pp. 8–9. (It also now forms a chapter in volume III of my *Postscript to the Logic of Scientific Discovery*, see my *Quantum Theory and the Schism in Physics*, edited by W.W. Bartley, III, Hutchinson, London, 1982.)

The fundamental idea (that the inertial mass of the electron is in part explicable as the inertia of the moving electromagnetic field) which led to the electromagnetic theory of matter is due to J.J. Thomson, 'On the Electric and Magnetic Effects produced by the Motion of Electrified Bodies', *Philosophical Magazine*, fifth series, *11*, 1881, pp. 229–49, and to O. Heaviside, 'On the Electromagnetic Effects due to the Motion of Electrification through a Dielectric', *Philosophical Magazine*, fifth series, *27*, 1889, pp. 324–39. It was developed by W. Kaufmann ('Die magnetische und elektrische Ablenkbarkeit der Bequerelstrahlen und die scheinbare Masse der Elektronen', *Gött. Nachr.*, 1901, pp. 143–55, 'Ueber die elektromagnetische Masse des Elektrons', 1902, pp. 291–6, 'Ueber die "Elektromagnetische Masse" der Elektronen', 1903, pp. 90–103) and M. Abraham ('Dynamik des Elektrons', *Gött. Nachr.*, 1902, pp. 20–41, 'Prinzipien der Dynamik des Elektrons', *Annalen der Physik*, fourth series, *10*, 1903, pp. 105–79) into the thesis that the mass of the electron is a purely electromagnetic effect. (See W. Kaufmann, 'Die elektromagnetische Masse des Elektrons', *Physikalische Zeitschrift*, *4*, 1902–3, pp. 57–63 and M. Abraham, *Theorie der Elektrizität*, volume II, Leipzig, 1905, pp. 136–249.) The idea was strongly supported by H.A. Lorentz, 'Elektromagnetische vershijnselen in een stelsel dat zich met willekeurige snelheid, kleiner dan die vanhet licht, beweegt', *Verslag van de Gewone Vergadering der Wis – en Natuurkundige Afdeeling, Koninklijke Akademie van Wetenschappen te Amsterdam*, XII, 1903–4, second part, pp. 986–1009, and by Einstein's special relativity, leading to results deviating from those of Kaufmann and Abraham. The electromagnetic theory of matter had a great ideological influence on scientists because of the fascinating possibility of *explaining* matter. It was shaken and modified by Rutherford's discovery of the nucleus (and the proton) and by Chadwick's discovery of the neutron, which may help to explain why its final overthrow by the theory of nuclear forces was hardly noticed.

46 The revolutionary power of special relativity lies in a new point of view which allows the derivation and interpretation of the Lorentz transformations from two simple first principles. The greatness of this revolution can be best gauged by reading Abraham's book (volume II, referred to in the preceding note). This book, which is slightly earlier than Poincaré's and Einstein's papers on relativity, contains a full discussion of the problem situation: of Lorentz's theory of the Michelson experiment, and even of Lorentz's local time. Abraham comes, e.g. on pp. 143f. and 370f., quite close to Einsteinian ideas. It even seems as if Max Abraham was better informed about the problem situation than was Einstein. Yet there is no realization of the revolutionary potentialities of the problem situation – quite the contrary. For Abraham writes in his Preface, dated March 1905: 'The theory of electricity now appears to have entered a state of quieter development.' This shows how hopeless it is even for a great scientist like Abraham to foresee the future development of his science.

47 See H. Minkowski, 'Space and Time', in A. Einstein, H.A. Lorentz, H. Weyl, and H. Minkowski, *The Principle of Relativity*, Methuen, London, 1923 and Dover edition, New York, p. 75.

48 Cornelius Lanczos, 'Rationalism and the Physical World', in *Boston Studies in the Philosophy of Science*, 3, edited by R.S. Cohen and M.W. Wartofsky, Reidel, Dordrecht, 1967, pp. 181–98. See p. 198.

49 See my *Conjectures and Refutations*, p. 114 (with footnote 30). See also my *Open Society*, volume II, p. 20, and the criticism in my *Logic of Scientific Discovery*, p. 440. I pointed out this criticism in 1950 to P.W. Bridgman, who received it most generously.

50 See A.S. Eddington, *Space, Time and Gravitation*, Cambridge University Press, Cambridge, 1935, pp. 162f. It is interesting in this context that Dirac (on p. 46 of the paper referred to in note 29 above) says that he now doubts whether four-dimensional thinking is a fundamental requirement of physics. (It is a fundamental requirement for driving a motor car.)

51 More precisely, a body falling from infinity with a velocity $v > c/3^{1/2}$ towards the centre of a gravitational field will be constantly decelerated in approaching this centre.

52 See the reference to Troels Eggers Hansen cited in note 27 above. See also Peter Havas, 'Four-Dimensional Formulations of Newtonian Mechanics and their Relation to the Special and the General Theory of Relativity', *Review of Modern Physics*, 36, 1964, pp. 938–65, and 'Foundation Problems in General Relativity', in *Delaware Seminar in the Foundations of Physics*, edited by M. Bunge, Springer-Verlag, New York, 1967, pp. 124–48. Of course, the comparison is not trivial: see for example pp. 52f. of E. Wigner's book referred to in note 24 above.

53 See Lanczos, 'Rationalism and the Physical World', p. 196.

54 See Leopold Infeld, *Quest*, Victor Gollancz, London, 1941, p. 90.

55 See Albert Einstein, 'Die Feldgleichungen der Gravitation', *Preussische Akademie der Wissenschaften, Sitzungsberichte*, 1915, pt. 2, pp. 844–7; 'Die Grundlage der allgemeinen Relativitätstheorie', *Annalen der Physik*, fourth series, *49*, 1916, pp. 769–822.

56 Thus I believe that §2 of Einstein's famous paper, 'Die Grundlage der allgemeinen Relativitätstheorie' (see note 55 above; English translation, 'The Foundation of the General Theory of Relativity', in *The Principle of Relativity*, pp. 111–64; see note 47 above) uses most questionable epistemological arguments *against* Newton's absolute space and *for* a very important theory.

57 Especially Heisenberg and Bohr.

58 Apparently it affected Max Delbrück. See Donald Fleming, 'Émigré Physicists and the Biological Revolution', *Perspectives in American History*, 2, Harvard, 1968, pp. 152–89, especially sections iv and v. (I owe this reference to Professor Mogens Blegvad.)

59 It is clear that a physical theory which does not explain such constants as the electric elementary quantum (or the fine structure constant) is incomplete – to say nothing of the mass spectra of the elementary particles. See my paper 'Quantum Mechanics without "The Observer"', referred to in note 45.

2

THE MYTH OF THE
FRAMEWORK

Those who believe this, and those who do not, have no
common ground of discussion, but in view of their opinions
they must of necessity scorn each other.

Plato

I

One of the more disturbing aspects of the intellectual life of our time
is the way in which irrationalism is so widely advocated, and the way
in which irrationalist doctrines are taken for granted. One of the
components of modern irrationalism is relativism (the doctrine that
truth is relative to our intellectual background, which is supposed to
determine somehow the framework within which we are able to
think: that truth may change from one framework to another), and,
in particular, the doctrine of the impossibility of mutual understanding between different cultures, generations, or historical
periods – even within science, even within physics. In this paper I
discuss the problem of relativism. It is my claim that behind it lies
what I call 'The Myth of the Framework'. I explain and criticize this
myth, and comment also on arguments that have been used in its
defence.

The proponents of relativism put before us standards of mutual
understanding which are unrealistically high. And when we fail to

Based on a paper which was first prepared in 1965, a version of which was first
published in *The Abdication of Philosophy: Philosophy and the Public Good* (The
Schilpp *Festschrift*), edited by E. Freeman, The Open Court Publishing Co., La Salle,
Illinois, 1976. I am indebted to Arne Petersen for various suggestions and corrections,
and to Alan Musgrave for reminding me to include the diagnosis contained in section
XVI. The motto is from Plato's *Crito*, 49D.

meet those standards, they claim that understanding is impossible. Against this, I argue that if common goodwill and a lot of effort are put into it, then very far-reaching understanding is possible. Furthermore, the effort is amply rewarded by what we learn in the process about our own views, as well as about those we are setting out to understand.

This paper sets out to challenge relativism in its widest sense. It is important to present such a challenge, for today the increasing escalation in the production of weapons has made survival almost identical with understanding.

II

Although I am an admirer of tradition, and conscious of its importance, I am, at the same time, an almost orthodox adherent of unorthodoxy: *I hold that orthodoxy is the death of knowledge, since the growth of knowledge depends entirely on the existence of disagreement.* Admittedly, disagreement *may* lead to strife, and even to violence. And this, I think, is very bad indeed, for I abhor violence. Yet disagreement may also lead to discussion, to argument, and to mutual criticism. And these, I think, are of paramount importance. I suggest that the greatest step towards a better and more peaceful world was taken when the war of swords was first supported, and later sometimes even replaced, by a war of words. This is why my topic is of some practical significance.

III

But let me first explain what my topic is, and what I mean by my title, 'The Myth of the Framework'. I will discuss, and argue against, a myth: a false story that is widely accepted, especially in Germany. From there it invaded America, where it became almost all-pervasive among intellectuals, and where it forms the background of some of the most flourishing schools of philosophy. So I fear that the majority of my present readers may also believe in the myth, either consciously or unconsciously.

The myth of the framework can be stated in one sentence, as follows.

A rational and fruitful discussion is impossible unless the participants share a common framework of basic assumptions or, at least, unless they have agreed on such a framework for the purpose of the discussion.

This is the myth I am going to criticize.

As I have formulated it here, the myth sounds like a sober statement, or like a sensible warning to which we ought to pay attention in order to further rational discussion. Some people even think that what I describe as a myth is a logical principle, or based on a logical principle. I think, on the contrary, that it is not only a false statement, but also a vicious statement which, if widely believed, must undermine the unity of mankind, and so must greatly increase the likelihood of violence and of war. This is the main reason why I want to combat it, and to refute it.

As indicated, I mean by 'framework' here a set of basic assumptions, or fundamental principles – that is to say, an *intellectual* framework. It is important to distinguish such a framework from some attitudes which may indeed be preconditions for a discussion, such as a wish to get to, or nearer to, the truth, and a willingness to share problems or to understand the aims and the problems of somebody else.

Let me say at once that the myth contains a kernel of truth. Although I contend that it is a most dangerous exaggeration to say that a fruitful discussion is *impossible* unless the participants share a common framework, I am very ready to admit that a discussion among participants who do not share a common framework may be *difficult*. A discussion will also be difficult if the frameworks have little in common. And it will be the easier the greater the overlap between the frameworks. Indeed, if the participants agree on all points, it may turn out to be an easy, smooth, and rational discussion – though perhaps a little boring.

But what about fruitfulness? In the formulation I gave of the myth, it is a *fruitful* discussion which is declared impossible. Against this I shall defend the directly opposite thesis: that a discussion between people who share many views is unlikely to be fruitful, even though it may be pleasant; while a discussion between vastly different frameworks can be extremely fruitful, even though it may sometimes be extremely difficult, and *perhaps* not quite so pleasant (though we may learn to enjoy it).

I think that we may say of a discussion that it was the more fruitful the more its participants were able to learn from it. And this means: the more interesting questions and difficult questions they were asked, the more new answers they were induced to think of, the more they were shaken in their opinions, and the more they could see things differently after the discussion – in short, the more their intellectual horizons were extended.

Fruitfulness in this sense will almost always depend on the original gap between the opinions of the participants in the discussion. The greater the gap, the more fruitful the discussion *can* be – always provided, of course, that such a discussion is not altogether *impossible*, as the myth of the framework asserts.

IV

But is a fruitful discussion between different frameworks really possible? Let us take an extreme case. Herodotus, the father of historiography, tells an interesting though somewhat gruesome story of the Persian King, Darius the First, who wanted to teach a lesson to the Greeks living in his empire. It was the custom of the Greeks to burn their dead. Darius 'summoned', we read in Herodotus, 'the Greeks living in his land, and asked them for what payment they would consent to eat up their fathers when they died. They answered that nothing on earth would induce them to do so. Then Darius summoned the ... Callatians, who do eat their fathers, and he asked them in the presence of the Greeks, who had the help of an interpreter, for what payment they would consent to burn the bodies of their fathers when they died. And they cried out aloud and implored him not to mention such an abomination.'[1]

Darius, I suspect, wanted to demonstrate the truth of something like the myth of the framework. Indeed, we are given to understand that a discussion between the two parties would have been impossible even with the help of that interpreter. It was an extreme case of a *'confrontation'* – to use a term much in vogue with believers in the myth of the framework, and a term which they like to use when they wish to draw our attention to the fact that a 'confrontation' rarely results in a fruitful discussion.

Let us assume that this confrontation staged by King Darius actually did take place as Herodotus narrates it. Was it really fruitless? I deny that it was. Admittedly, it does not seem that mutual understanding was achieved. And the story shows that we *may* be faced, in some rare cases, by an unbridgeable gulf. But even in this case, there can be little doubt that both parties were deeply shaken by the experience, and that they learned something new. I myself find the idea of cannibalism just as revolting as did the Greeks at the court of King Darius. And I suppose my readers will feel the same. But these feelings should make us all the more perceptive and the more appreciative of the admirable lesson which Herodotus wishes us to draw from the story. Alluding to Pindar's distinction between nature and convention,[2] Herodotus suggests that we should look with tolerance and even with respect upon customs or conventional laws that differ from our own. If this particular confrontation ever took place, some of the participants may well have reacted to it in the enlightened way in which Herodotus wishes us to react to his story.

This shows that there is, even without a discussion, a possibility of a fruitful confrontation among people deeply committed to different frameworks. *But we must not expect too much*: we must *not* expect that a confrontation, or even a prolonged discussion, will end with the participants reaching *agreement*.

But is agreement *always* desirable? Let us assume that there is a discussion and that the issue at stake is the truth or falsity of some theory or hypothesis. We – that is, the rational witnesses or, if you like, impartial spectators of the discussion – should of course like the discussion to end with all parties agreeing that the theory is true *if in fact* it *is true*, or that the theory is false *if in fact* it *is false*: but *only* in these cases. For we should like the discussion to reach, if possible, a true verdict. We should, however, dislike the idea that agreement was reached on the truth of the theory if the theory was in fact false. And even if it was true, we should prefer that no agreement is reached on its truth if the arguments supporting the theory were too weak to bear out the conclusion. In such a case, we might even prefer that no agreement is reached at all. And in such a case we should say that the discussion was fruitful if the clash of opinion led the participants to produce new and interesting arguments, even though these arguments were inconclusive. For conclusive arguments in support of a theory are very rare in all but the most trivial issues, even though arguments against a theory may sometimes be pretty strong.

Looking back at Herodotus' story of the confrontation, we can now see that even in this extreme case where no agreement was in sight the confrontation may have been useful and that, given time and patience (which Herodotus seems to have had at his disposal), it did bear fruit – at least in Herodotus' own mind.

Thus my thesis is not that the gulf between different frameworks, or between different cultures, can, for logical reasons, *always* be bridged. My thesis is merely that it can *usually* be bridged. There may be no common assumptions. There may perhaps be only common problems. For different groups of humans do, as a rule, have much in common, such as the problems of survival. But even common problems may not always be needed. My thesis is that logic neither underpins the myth of the framework nor its denial, but that we can try to learn from each other. Whether we succeed will depend largely on our goodwill, and to some extent also on our historical situation, and on our problem situation.

V

I wish to suggest here that, in a way, we ourselves and our own attitudes are in part the results of confrontations and of inconclusive discussions of the kind described by Herodotus.

What I mean can be summed up by the thesis that our Western civilization is the result of the clash, or confrontation, of different cultures, and therefore of the clash, or confrontation, of different frameworks.

It is widely admitted that our civilization – which at its best may be described, somewhat eulogistically, as a rationalist civilization – is very largely the result of the Greco-Roman civilization. This civilization acquired many of its features, such as the alphabet, even before the clashes between the Romans and the Greeks, through its clashes with the Egyptian, the Persian, the Phoenician, and other Middle Eastern civilizations. And in the Christian era our civilization was further modified through clashes with the Jewish civilization and through clashes due to the Germanic and the Islamic invasions.

But what of the original Greek miracle – the rise of Greek poetry, art, philosophy, and science: the real origin of Western rationalism? I assert that the Greek miracle, *so far as it can be explained*, was also largely due to culture clash. It seems to me that this is indeed one of the lessons which Herodotus wants to teach us in his *History*.

Let us look for a moment at the origin of Greek philosophy and Greek science.[3] It all began in the Greek colonies: in Asia Minor, in Southern Italy, and in Sicily. These are places where the Greek colonists were confronted with the great civilizations of the East, and clashed with them, or where, in the West, they met Sicilians, Carthaginians, and Italians such as the Tuscans. The impact of culture clash on Greek philosophy is very obvious from the earliest reports about Thales, the founder of Greek philosophy. It is unmistakable in Heraclitus, who seems to have been influenced by Zoroaster. But the way in which culture clash may lead men to think critically comes out most forcefully in Xenophanes, the wandering bard. Although I have quoted some of his verses on other occasions, I will do so again, because they illustrate this point so beautifully.[4] Xenophanes makes use of the lessons he learned from the clash between the Greek, the Ethiopian, and the Thracian cultures for a criticism of the anthropomorphic theologies of Homer and Hesiod:

> The Ethiops say that their gods are flat-nosed and black
> While the Thracians say that theirs have blue eyes and red hair.
> Yet if cattle or horses or lions had hands and could draw
> And could sculpture like men, then the horses would draw their
>   gods
> Like horses, and cattle like cattle, and each would then shape
> Bodies of gods in the likeness, each kind, of its own.

And Xenophanes draws an important critical conclusion from this lesson: he concludes that human knowledge is fallible:

> The gods did not reveal, from the beginning,
> All things to us; but in the course of time,
> Through seeking we may learn, and know things better ...
> These things are, we conjecture, like the truth.

> But as for certain truth, no man has known it,
> Nor will he know it; neither of the gods,
> Nor yet of all the things of which I speak.
> And even if by chance he were to utter
> The final truth, he would himself not know it:
> For all is but a woven web of guesses.

Although Burnet and others have denied it, I think that Parmenides, perhaps the greatest of these early thinkers, stood under Xenophanes' influence.[5] He takes up Xenophanes' distinction between the one final truth which is not subject to human convention, and the guesses, opinions, and conventions of the mortals. There are always many different opinions and conventions concerning any one problem or subject-matter (such as the gods). This shows that they are not all true. For if they conflict, then at best only one of them can be true.[6] Thus it appears that Parmenides (a contemporary of Pindar to whom Plato attributes the distinction between nature

39

and convention) was the first to distinguish clearly between truth or reality on the one hand, and convention or conventional opinion (hearsay, plausible myth) on the other – a lesson which, we may say, he derived from Xenophanes and from culture clash. It led him to one of the boldest theories ever conceived.

Culture clash played an important role in the rise of Greek science – of mathematics and astronomy – and one can even specify the way in which some of the various clashes bore fruit. Moreover, our ideas of freedom, of democracy, of toleration, and also the ideas of knowledge, of science, of rationality, can all be traced back to these early Greek experiences.

Of all these ideas, the idea of rationality seems to me the most fundamental. It appears from our sources that the invention of rational or critical discussion was contemporaneous with some of these clashes, and that discussion became traditional with the rise of the earliest Ionian democracies.

VI

One of the main tasks for human reason is to make the universe we live in understandable to ourselves. This is the task of science. There are two different components of about equal importance in this enterprise.

The first is poetic inventiveness, that is, story-telling or myth-making: the invention of stories which explain the world. These are, to begin with, often or perhaps always polytheistic. Men feel that they are in the hands of *unknown powers*, and they try to understand and to explain the world, and human life and death, by inventing stories or myths about these powers.

This first component, which may be perhaps as old as human language itself, is all-important. And it seems universal: all tribes, and all peoples, have such explanatory stories, often in the form of fairy tales. It seems that the invention of explanations and explanatory stories is one of the basic functions that human language has to serve.

The second component of rationality is of comparatively recent date. It seems to be specifically Greek and to have arisen after the establishment of writing in Greece. It arose, it seems, only once: with Anaximander, the pupil of Thales, and the first critical cosmologist. It is the invention of criticism, of the critical discussion of the various explanatory myths – with the aim of consciously improving upon them.

The main Greek example of explanatory myth-making on an elaborate scale is, of course, Hesiod's *Theogony*. This is a wild and gruesome story of the origin, the deeds, and the misdeeds, of the Greek gods. At first sight one would hardly feel inclined to think that the *Theogony* may provide suggestions which could influence the development of a scientific explanation of our world. Yet I have proposed the historical conjecture that a passage in Hesiod's *Theogony*[7], which was foreshadowed by another in Homer's *Iliad*[8], was so used by Anaximander.

I will explain my conjecture. According to tradition, Thales – the teacher and kinsman of Anaximander, and the founder of the Ionian school of cosmologists – taught that 'the earth is supported by water on which it rides like a ship'. Anaximander – the eventual successor of Thales – turned away from this somewhat naïve myth (intended by Thales to explain earthquakes). Anaximander's new departure was of a truly revolutionary character, for he taught, we are told, the following: 'There is no thing at all that is holding up the earth. Instead, the earth remains stationary owing to the fact that it is equally far away from all other things. Its shape is like that of a drum. We walk on one of its flat surfaces while the other is on the opposite side.'

This bold idea made possible the ideas of Aristarchus and Copernicus, and it even contains an anticipation of Newton's forces. How did it arise? I have also proposed the conjecture[9] that it arose out of a purely logical criticism of Thales' myth. The criticism is simple: if we explain the position and stability of the earth in the universe by saying that it is supported by the ocean like a ship that is supported by water, then we are bound, the critic says, to explain the position and stability of the ocean. But this would mean finding some support for the ocean, and then some further support for this support. Obviously, this leads to an infinite regress. How can we avoid it?

In looking for a way out of this frightful impasse which, it appeared, no alternative explanation was able to avoid, Anaximander remembered, I conjecture, a passage in which Hesiod develops an idea from the *Iliad*, where we are told that Tartarus is exactly as far beneath the earth as Uranus, or heaven, is above it.

The passage reads: 'For nine days and nights will a brazen anvil fall from the heavens, and on the tenth it will reach the earth. And for nine days and nights will a brazen anvil fall from the earth, and on the tenth it will reach Tartarus.'[10] This passage may have suggested to Anaximander that we can draw a diagram of the world, with the earth in the middle, and the vault of the heavens like a hemisphere above it. Symmetry then suggests that we interpret Tartarus as being the lower half of the vault. In this way we arrive at Anaximander's construction as it is transmitted to us – a construction that breaks through the deadlock of the infinite regress.

There is, I think, a need for such a conjectural explanation of the tremendous step that carried Anaximander beyond his teacher Thales. My conjecture, it seems to me, makes the step more understandable and, at the same time, even more impressive: for it is now seen as a rational solution of a very difficult problem – the problem of the support and the stability of the earth.

Yet Anaximander's criticism of Thales and his critical construction of a new myth would have led to nothing had these not been followed up. How can we explain the fact that they *were* followed up? Why was a new myth offered in each generation after Thales? I have tried to explain this by the further conjecture that Thales and Anaximander together founded a new school tradition – *the critical tradition*.

My attempt to explain the phenomenon of Greek rationalism and of the Greek critical tradition by a school tradition is, of course, again completely conjectural. In fact, it is itself a kind of myth. Yet it does explain a unique phenomenon – the Ionian school. For at least four or five generations this school produced, in each new generation, an ingenious revision of the teachings of the preceding generation. In the end it established what we may call the scientific tradition: a tradition of criticism which survived for at least five hundred years, and which resisted some serious onslaughts before it succumbed.

The critical tradition was founded by the adoption of the method of criticizing a received story or explanation and then proceeding to a new, improved, imaginative story which in turn is submitted to criticism. This method, I suggest, is the method of science. It seems to have been invented only once in human history. It died in the West when the schools in Athens were suppressed by a victorious and intolerant Christianity, though it lingered on in the Arab East. It was missed and mourned during the Middle Ages. In the Renaissance, it was not so much reinvented as reimported from the

East, together with the rediscovery of Greek philosophy and Greek science.

The uniqueness of this second component of the scientific tradition – the method of critical discussion – will be realized if we consider the old-established function of schools, especially of religious and semi-religious schools. Their function is, and has always been, to preserve the purity of the teaching of the founder of the school. Accordingly, changes in doctrine are rare and are mainly due to mistakes or misunderstandings. When they are consciously made they are made, as a rule, surreptitiously – for otherwise changes will lead to splits, to schisms.

Yet in the Ionian school, we find a school tradition which carefully preserved the teaching of each of its masters, while it nonetheless deviated from their teaching afresh in each new generation.

My conjectural explanation of this unique phenomenon is that it started when Thales, the founder, encouraged Anaximander, his follower, to see whether he could produce a better explanation of the apparent stability of the earth than he himself had been able to offer.

However this may have been, the invention of the critical method could hardly have happened without the impact of culture clash. And it had the most tremendous consequences. For within four or five generations, Parmenides boldly suggested that the earth, the moon, and the sun were spheres, that the moon moved round the earth while she was always 'wistfully' looking 'round for the rays of the sun', and that this could be explained by the assumption that she borrowed her light from the sun.[11] A little later it was conjectured in the Platonic school that the earth rotated, and that the earth moved round the sun. But these later hypotheses, due especially to Aristarchus, seem to have been too bold, and they were soon forgotten.

These cosmological or astronomical findings became the basis of all future science. Human science started from a bold and hopeful attempt to understand critically the world in which we live. This ancient dream found fulfilment in Newton. We can say that only since Newton has humanity become fully conscious – conscious of its position in the universe.

All this, I suggest, is the result of culture clash, or the clash of frameworks, which led to the application of the method of critical discussion to myth-making – to our attempts to understand, and to explain the world to ourselves.

VII

If we look back on this development, then we can understand better why we must not expect any critical discussion of a serious issue, any 'confrontation', to yield quick and final results. Truth is hard to come by. It needs both ingenuity in criticizing old theories, and ingenuity in the imaginative invention of new theories. This is so not only in the sciences, but in all fields.

Serious critical discussions are always difficult. Non-rational human elements such as personal problems always enter. Many participants in a rational, that is, a critical, discussion find it particularly difficult that they have to unlearn what their instincts seem to teach them (and what they are taught, incidentally, by every debating society): that is, to win. For what they have to learn is that victory in a debate is nothing, while even the slightest clarification of one's problem – even the smallest contribution made towards a clearer understanding of one's own position or that of one's opponent – is a great success. A discussion which you win but which fails to help you to change or to clarify your mind at least a little should be regarded as a sheer loss. For this very reason no change in one's position should be made surreptitiously, but it should always be stressed and its consequences explored.

Rational discussion in this sense is a rare thing. But it is an important ideal, and we may learn to enjoy it. It does not aim at conversion, and it is modest in its expectations: it is enough, more than enough, if we feel that we can see things in a new light or that we have got even a little nearer to the truth.

VIII

But let me now return to the myth of the framework. There are many tendencies which may contribute to the fact that this myth is often taken for an almost self-evident truth.

One of these tendencies I have already mentioned. It results from a disappointed over-optimism concerning the powers of reason – that is, from an over-optimistic expectation concerning the outcome of a discussion. I mean the expectation that discussion should lead to a decisive and deserved intellectual victory of the truth, represented by one party, over falsity, represented by the other. When it is found that this is not what a discussion usually achieves, disappointment turns an over-optimistic expectation into a general pessimism concerning the fruitfulness of discussions.

Another tendency which contributes to the myth of the framework, and which deserves careful scrutiny, is connected with historical or cultural relativism. This is a view whose beginnings may perhaps be discerned in Herodotus.

Herodotus seems to have been one of those somewhat rare people whose minds are broadened by travel. At first he was no doubt shocked by the many strange customs and institutions which he encountered in the Middle East. But he learned to respect them, and to look on some of them critically, and to regard others as the results of historical accidents: he learned to be tolerant, and he even acquired the ability to see the customs and institutions of his own country through the eyes of his barbarian hosts.

This is a healthy attitude. But it may lead to relativism, that is, to the view that there is no absolute or objective truth, but rather one truth for the Greeks, another for the Egyptians, still another for the Syrians, and so on.

I do not think that Herodotus fell into this trap. But many have done so since – perhaps inspired by an admirable feeling of tolerance, combined with some dubious logic.

There is one version of the idea of cultural relativism which is obviously correct. In England, Australia, and New Zealand we drive on the left hand side of the road, while in America, in Europe, and in most other countries we drive on the right hand side. What is clearly needed is *some* such rule of the road. But which of the two – the right or the left – is obviously arbitrary and conventional. There are many similar rules, of greater or lesser importance, which are purely conventional, and even arbitrary.[12] Among these are the different rules for pronouncing and spelling the English language in America and England. Even two quite different vocabularies could be related in such a purely conventional way. And provided the grammatical structures of the two languages are very similar, the situation would closely resemble that of the two different rules of the road. We may regard such vocabularies, or such rules, as differing in a purely conventional way: there is really nothing to choose between them – nothing, at least, of importance.

As long as we consider only conventional rules and customs such as these, there is no danger that the myth of the framework will be taken too seriously. For a discussion between an American and an Englishman about the rule of the road is likely to lead to an agreement. Both are likely to regret the fact that their rules do not coincide. Both will agree that in principle there is nothing to choose between the two rules and that it would be unreasonable to expect the United States to adopt the left hand rule in order to achieve conformity with Britain. And both are likely to agree that Britain cannot at present make a change which may be desirable but which would be extremely costly. After agreement has thus been reached on all points, both participants are likely to part with the feeling that they have learned nothing from the discussion which they did not already know.

The situation changes completely when we consider other institutions, laws, and customs – those, for example, which are connected with the administration of justice. Different laws and customs in this field may make all the difference for those living under them. Some laws and customs can be very cruel, while others provide for mutual help and the relief of suffering. Some countries and their laws respect freedom while others do so less, or not at all. These differences are most important, and they must not be dismissed or shrugged off by a cultural relativism, or by the claim that different laws and customs are due to different standards, or different ways of thinking, or different conceptual frameworks, and that they are therefore incommensurable or incomparable. On the contrary, we should try to understand and to compare. We should try to find out who has the better institutions. And we should try to learn from them.

It is my opinion that a critical discussion of these important matters is not only possible, but most urgently needed. It is often made difficult by propaganda and by a neglect of factual information. But these difficulties are not insuperable. Thus it is possible to combat propaganda by information. For information, if available, is not always ignored – though it is, admittedly, often ignored.

Cultural relativism and the doctrine of the closed framework are serious obstacles to the readiness to learn from others. They are obstacles to the method of accepting some institutions, modifying others, and rejecting what is bad. For instance, many people think that we can accept or reject only the whole framework or 'system' of 'communism' or of 'capitalism'. If we think about these so-called 'systems', we have to distinguish between the systems of theories – the ideologies – and certain social realities. Both have considerably influenced each other. But the social realities have little similarity to the ideologies – to what they are supposed to be, especially by Marxists.

IX

There are some people who uphold the myth that frameworks of laws and customs cannot be rationally discussed. They assert that morality is identical with legality or custom or usage, and that it is therefore impossible to judge, or even to discuss, whether one system of customs is morally better than another, since the various existing systems of laws and customs are the only possible standards of morality.

This view goes back to Hegel's famous formulae: 'What is real is reasonable' and 'What is reasonable is real'.[13] Here 'what is' or 'what is real' means the world, including its man-made laws and customs. That these are man-made is denied by Hegel, who asserted that the World Spirit or Reason made them, and that those who seem to have made them – the great men, the makers of history – are merely the executors of reason, their passions being the most sensitive instrument of reason. They are the detectors of the Spirit of their Time, and ultimately of the Absolute Spirit, that is of God himself.

(This is just one of those many cases in which philosophers use God for their own private purposes – that is, as a prop for some of their tottering arguments.)

Hegel was both a relativist and an absolutist: as always, he had it at least both ways, and if two ways were not enough, he had it in three ways. And he was the first of a long chain of post-Kantian, that is, post-critical or post-rationalist philosophers – mainly German – who upheld the myth of the framework.

According to Hegel, truth itself was both relative and absolute. It was relative to each historical and cultural framework: there could thus be no rational discussion between such frameworks since each of them had a different standard of truth. But Hegel held his own doctrine that truth was relative to the various frameworks to be absolutely true, since it was part of his own relativistic philosophy.

Hegel's claim to have discovered absolute truth does not now appear to attract many people. But his doctrine of relative truth and his version of the myth of the framework attracts them still. The most influential contributor to the myth after Hegel was, no doubt, Karl Marx. I need hardly remind you of his idea of a classbound science – of proletarian science and bourgeois science – each imprisoned within its own framework. After Marx, these ideas were further developed, especially by the German Max Scheler, and by the Hungarian Karl Mannheim. They called their theories 'the sociology of knowledge', and they held, like Marx, that each man's conceptual framework is determined by his 'social habitat'. I have criticized these ideas elsewhere, but it is important to see what is behind their appeal. What makes these ideas attractive is that people confuse relativism with the true and important insight that all men are fallible, and prone to be biased. This doctrine of fallibility has played a significant role in the history of philosophy from its earliest days on – from Xenophanes and Socrates to Erasmus and to Charles Sanders Peirce – and I think it is of the utmost importance. But I do *not* think that the true and important doctrine of human fallibility can be used to support relativism with respect to truth.

Of course, the doctrine of human fallibility can be validly used to argue against that kind of philosophical absolutism which claims that it *possesses* the absolute truth – or perhaps a *criterion* of absolute truth, such as the Cartesian criterion of clarity and distinctness, or some other intuitive criterion. But there exists a very different attitude towards absolute truth, in fact a fallibilist attitude. It stresses the fact that the mistakes we make can be absolute mistakes, in the sense that our theories can be absolutely false – that they can fall short of the truth. Thus to the fallibilist the notion of truth, and that of falling short of the truth, may represent absolute standards – even though we can never be certain that we are living up to them. But since they may serve as a kind of steering compass, they may be of decisive help in critical discussions.

This theory of absolute or objective truth has been revived by Alfred Tarski, who also proved that there can be no universal criterion of truth. There is no clash whatever between Tarski's theory of absolute or objective truth and the doctrine of fallibility.[14]

But is not Tarski's notion of truth a relative notion? Is it not relative to the language to which the statement whose truth is being discussed belongs?

The answer to this question is 'no'. Tarski's theory says that a statement in some language, say English, is true if and only if it corresponds to the facts. And Tarski's theory implies that whenever there is another language, say French, in which we can describe the same fact, then the French statement which describes this fact will be true if and only if the corresponding English statement is true. Thus it is impossible, according to Tarski's theory, that of two statements which are translations of each other, one can be true and the other false. Truth, according to Tarski's theory, is therefore *not* dependent on language, or relative to language. Reference to the language is made only because of the unlikely but trivial possibility that the same sounds or symbols may occur in two different languages and may then perhaps describe two totally different facts.

X

An awareness of the difficulties of translation between different languages has also contributed to the myth. It *may* happen that a statement in one language is untranslatable into another, or in other words that a fact or a state of affairs which can be described in one language cannot be described in another.

Anybody who can speak more than one language knows, of course, that perfect translations from one language into another are very rare, if they exist at all. But this difficulty, well known to all translators, should be clearly distinguished from the situation discussed here – that is, the impossibility of describing in one language a state of affairs which can be described in some other language. The ordinary and well-known difficulty consists of something quite different, namely this. A crisp, simple, and easily understandable statement in French or English may need a highly complex and awkward rendering in, say, German – a rendering which is even difficult to understand in German. In other words, the ordinary difficulty known to every translator is that an aesthetically adequate translation may be impossible, not that *any* translation of the statement in question is impossible. (I am speaking here of a factual statement, not of a poem or an aphorism or *bon mot*, or of a statement which is subtly ironical or which expresses a sentiment of the speaker.)

There can be no doubt, however, that a more radical impossibility may arise. For example, we can construct an artificial language which contains only one-termed predicates, so that we can say in this language 'Paul is tall' and 'Peter is short', but not 'Paul is taller than Peter'.

More interesting than such artificial languages are some living languages. Here we can learn much from Benjamin Lee Whorf.[15] Whorf was, it seems, the first to draw attention to the significance of certain tenses in the language of the Hopi, an American Indian tribe. These tenses are experienced by a Hopi speaker as describing some part of the state of affairs which he tries to describe in his statement. They cannot be adequately rendered into English, for we can explain them only in a roundabout way, by referring to certain expectations of the speaker rather than aspects of the objective states of affairs.

Whorf gives the following example. There are two tenses in Hopi which might inadequately be rendered in English by the two statements: 'Fred began chopping wood' and 'Fred began to chop wood'. The first would be used by the Hopi speaker if he expects Fred to *go on* chopping for some time. If the speaker does *not* expect Fred to go on chopping, then he will *not* say, in Hopi, 'Fred began chopping'. He will use that other tense rendered by 'Fred began to chop'. But the real point is that the Hopi speaker does not wish by the use of his tenses merely to express his different expectations. He rather wishes to describe two different states of affairs – two different objective situations, two different states of the objective world. The one tense may be said to describe the beginning of a continuing *state* or of a somewhat repetitive *process*, while the other describes the beginning of an *event* of short duration. Thus the Hopi speaker may try to translate Hopi into English by saying: 'Fred began sleeping', in contradistinction to 'Fred began to sleep', because sleeping is a process rather than an event.

All this is very much simplified: a full restatement of Whorf's description of the complex linguistic situation could easily take up a whole paper. The main consequence for my topic which seems to emerge from the situations described by Whorf and more recently discussed by Quine is this. Although there cannot be any linguistic relativity concerning the *truth* of any statement, there is the possibility that a statement may be untranslatable into certain other languages. For two different languages may have built into their very grammar two different views of the stuff the world is made of, or of the world's basic structural characteristics. In the terminology of Quine this may be called the 'ontological relativity' of language.[16]

The possibility that some statements are untranslatable is, I suggest, about the most radical consequence we can draw from what Quine calls 'ontological relativity'. Yet in spite of Quine's various striking but somewhat *a priori* arguments against translatability, most human languages are in fact reasonably inter-translatable. Of course, some languages are *badly* inter-translatable – perhaps because of ontological relativity, perhaps for other reasons.[17] For example, an appeal to our sense of humour, or an allusion to a locally well-known historical event, may be completely untranslatable.

XI

It is obvious that this situation must make rational discussion very difficult if the participants are brought up in different parts of the world and speak different languages. But I have found that these difficulties can often be surmounted. I have had students in the London School of Economics not only from Europe and America but also from various parts of Africa, the Middle East, India, South-East Asia, China, and Japan. And I have found that the difficulties could usually be conquered with a little patience on both sides. Whenever there was a major obstacle to overcome, it was, as a rule, the result of indoctrination with Western ideas. Dogmatic, uncritical teaching in bad Westernized schools and universities, and especially training in Western verbosity and in some Western ideology were, in my experience, much graver obstacles to rational discussion than any cultural or linguistic gap.

These experiences also suggested to me that culture clash may lose some of its great value if one of the clashing cultures regards itself as universally superior, and even more so if it is so regarded by the other: this may destroy the greatest value of culture clash, for the greatest value of culture clash lies in the fact that it can evoke a critical attitude. More especially, if one of the parties becomes convinced of its inferiority, then the critical attitude of trying to learn from the other will be replaced by a kind of blind acceptance: a blind leap into a new magic circle, or a conversion, as it is so often described by fideist and existentialist philosophers.

I believe that ontological relativity, though an obstacle to easy communication, can prove of immense value in all the more important cases of culture clash if it can be overcome not by a sudden leap into the dark, but sufficiently slowly. For it means that the partners in the clash may liberate themselves from prejudices of which they are unconscious – from taking theories unconsciously for granted, theories which, for example, may be embedded in the logical structure of their language. Such a liberation may be the result of criticism awakened by culture clash.

What happens in such cases? We compare and contrast the new language with our own, or with some others we know well. In the comparative study of these languages we use, as a rule, our own language as a metalanguage – that is, as the language in which we speak about, and compare, the languages which are the objects under investigation, including our own language. The languages under

51

investigation are the object languages. In carrying out the investigation, we are forced to look upon our own language – say English – in a critical way, as a set of rules and usages which may be somewhat narrow since they are unable completely to capture, or to describe, the kinds of entities which the other languages assume to exist. But this description of the limitations of English as an object language is carried out in English as a metalanguage. Thus we are forced by this comparative study to transcend precisely those limitations which we are studying. And the interesting point is that we succeed in this. The means of transcending our language is *criticism*.

Whorf himself, and some of his followers, have suggested that we live in a kind of intellectual prison, a prison formed by the structural rules of our language. I am prepared to accept this metaphor, though I have to add to it that it is an odd prison as we are normally unaware of being imprisoned. We may become aware of it through culture clash. But then, this very awareness allows us to break out of the prison. If we try hard enough, we can transcend our prison by studying the new language and by comparing it with our own.

Admittedly, the result will be a new prison. But it will be a much larger and wider prison. And again, we will not suffer from it. Or rather, whenever we do suffer from it, we are free to examine it critically, and thus to break out again into a still wider prison.

The prisons are the frameworks. And those who do not like prisons will be opposed to the myth of the framework. They will welcome a discussion with a partner who comes from another world, from another framework, for it gives them an opportunity to discover their so far unfelt chains, to break these chains, and thus to transcend themselves. But this breaking out of one's prison is clearly not a matter of routine:[18] it can only be the result of a critical effort and of a creative effort.

XII

In the remainder of this paper I will try to apply this brief analysis to some problems which have arisen in a field in which I am greatly interested – the philosophy of science.

It is now fifty years since I arrived at a view very similar to the myth of the framework – and I not only arrived at it but at once went beyond it. It was during the great and heated discussions after the First World War that I found out how difficult it was to get anywhere with people living in a closed framework – I mean people like the Marxists, the Freudians, and the Adlerians. None of them could ever be shaken in his adopted view of the world. Every argument against their framework was interpreted by them so as to fit into it. And if this turned out to be difficult, then it was always possible to psychoanalyse or socioanalyse the arguer: criticism of Marxian ideas was due to class prejudice, criticism of Freudian ideas was due to repression, and criticism of Adlerian ideas was due to the urge to prove your superiority, an urge which was due to an attempt to compensate for a feeling of inferiority.

I found the stereotyped pattern of these attitudes depressing and repelling, the more so as I could find nothing of the kind in the debates of the physicists about Einstein's General Theory, although it too was hotly debated at the time.

The lesson I derived from these experiences was this. Theories are important and indispensable because without them we could not orientate ourselves in the world – we could not live. Even our observations are interpreted with their help. The Marxist literally *sees* class struggle everywhere. Thus he believes that only those who deliberately shut their eyes can fail to see it. The Freudian sees everywhere repression and sublimation. The Adlerian sees how feelings of inferiority express themselves in every action and every utterance, whether it is an utterance of inferiority or superiority.

This shows that our need for theories is immense, and so is the power of theories. Thus it is all the more important to guard against becoming addicted to any particular theory: we must not let ourselves be caught in a mental prison. I did not know of the theory of culture clash at the time, but I certainly made use of my clashes with the addicts of the various frameworks in order to impress upon my mind the ideal of liberating oneself from the intellectual prison of a theory in which one might get stuck unconsciously at any moment of one's life.

It is only too obvious that this idea of self-liberation, of breaking out of one's prison of the moment, might in its turn become part of a framework or a prison – or in other words, that we can never be absolutely free. But we can widen our prison, and at least we can leave behind the narrowness of one who is addicted to his fetters.

Thus our view of the world is at any moment necessarily theory-impregnated. But this does not prevent us from progressing to better theories. How do we do it? The essential step is the linguistic formulation of our beliefs. This objectifies them, and thus makes it possible for them to become targets of criticism. In this way, our beliefs are replaced by competing theories, by competing conjectures. And through the critical discussion of these theories we can progress.

In this way we must demand of any better theory, that is, of any theory which may be regarded as progressing beyond some less good theory, that it can be compared with the latter. In other words, that the two theories are *not* 'incommensurable', to use a now fashionable term, introduced in this context by Thomas Kuhn.

For example, Ptolemy's astronomy is far from incommensurable with that of Aristarchus and Copernicus. No doubt, the Copernican system allows us to see the world in a totally different way. No doubt there is, psychologically, a *Gestalt* shift, as Kuhn calls it. This is psychologically very important. But we *can* compare the two systems logically. In fact, it was one of Copernicus' main arguments that all astronomical observations which can be fitted into a geocentric system can, by a simple translation method, always be fitted into a heliocentric one. There is no doubt all the difference in the world between these two views of the universe, and the magnitude of the gulf between the two views may well make us tremble. But there is no difficulty in comparing them. For example, we may point out the colossal velocities which the rotating sphere of the fixed stars must give to the stars which are near to its equator, while the rotation of the earth, which in Copernicus' system replaces that of the fixed stars, involves very much smaller velocities. This, together with some practical acquaintance with centrifugal forces, may well have served as an important point of comparison for those who had to choose between the two systems.

I assert that this kind of comparison between systems which have historically grown out of the same problems (say, to explain the movements of the heavenly bodies) is always possible. Theories which offer solutions of the same or closely related problems are as a rule comparable, I assert, and discussions between them are always possible and fruitful. And not only are they possible, they actually take place.

XIII

Some people do not think that these assertions are correct, and this results in a view of science and its history very different from mine. Let me briefly outline such a view of science.

The proponents[19] of such a view observe that scientists are normally engaged in close cooperation and discussion. And the proponents argue that this situation is made possible by the fact that scientists normally operate within a common framework to which they have committed themselves. (Frameworks of this kind seem to me to be closely related to what Karl Mannheim used to call 'Total Ideologies'.[20]) The periods during which scientists remain committed to a framework are regarded as typical. They are periods of 'normal science', and scientists who work in this way are regarded as 'normal scientists'.

Science in this sense is then contrasted with science in a period of crisis or revolution. These are periods in which the theoretical framework begins to crack, and in the end breaks. It is then replaced by a new one. The transition from an old framework to a new one is regarded as a process which must be studied not from a logical point of view (for it is, essentially, not wholly, or even mainly, rational) but from a psychological and sociological point of view. There is, perhaps, something like 'progress' in the transition to a new theoretical framework. But this is not a progress which consists of getting nearer to the truth, and the transition is not guided by a rational discussion of the relative merits of the competing theories. *It cannot be so guided since a genuinely rational discussion is thought to be impossible without an established framework.* Without a framework it is not even thought to be possible to agree what constitutes a point of 'merit' in a theory. (Some protagonists of this view even think that we can speak of truth only relative to a framework.) Rational discussion is thus impossible if it is the framework which is being challenged. And this is why the two frameworks – the old and the new – have sometimes been described as *incommensurable*.

An additional reason why frameworks are sometimes said to be incommensurable seems to be this. A framework can be thought of as consisting not only of a 'dominant theory', but also as being, in part, a psychological and sociological entity. It consists of a dominant theory *together* with what one might call *a way of viewing things in tune with the dominant theory*, including sometimes even a way of viewing the world and a way of life. Accordingly, such a framework constitutes a social bond between its devotees: it binds them together, very much as a church does, or a political or artistic creed, or an ideology.

This is a further explanation of the asserted incommensurability: it is understandable that *two ways of life and two ways of looking at the world* are incommensurable. Yet I want to stress that *two theories* which try to solve the same family of problems, including their offspring (their problem children), *need* not be incommensurable, and that in science, as opposed to religion, it is *problems and theories* that are paramount. I do not wish to deny that there is such a thing as a 'scientific approach', or a scientific 'way of life' – that is, the way of life of those men devoted to science. On the contrary, I assert that the scientific way of life involves a burning interest in objective scientific theories – in the theories in themselves, and in the problem of their truth, or their nearness to truth. And this interest is a *critical* interest, an *argumentative* interest. Thus it does not, like some other creeds, produce anything like the described 'incommensurability'.

It seems to me that many counterexamples exist to the theory of the history of science that I have just discussed. There are, first, counterexamples that show that the existence of a 'framework' – of a dominant theory, and of work going on within it – is by no means a necessary or even a characteristic condition for the development of science. There are, more especially, counterexamples that show that there may be *several 'dominant' theories* struggling for centuries for supremacy in a science, and that there may even be fruitful discussions between them. My main counterexample under this heading is the theory of the constitution of matter, in which atomism and continuity theories were, fruitfully, at war from the Pythagoreans and Parmenides, Democritus and Plato, to Heisenberg and Schrödinger. I do not think that this war can be described as falling into the prehistory of science, or into the history of pre-science. Another counterexample of this second kind is constituted by the theories[21] of heat warring with kinetic and phenomenological theories. And the clash between Ernst Mach and Max Planck[22] was neither characteristic of a crisis, nor did it occur within one framework, nor, indeed, could it be described as pre-scientific. Another example is the clash between Cantor and his critics (especially Kronecker) which was later continued in the form of exchanges between Russell and Poincaré, Hilbert and Brouwer. By 1925 there were at least three sharply opposed frameworks involved, and they slowly changed their character. By now not only have fruitful discussions occurred but so many syntheses that the animadversions of the past are almost forgotten. Thirdly, there are counterexamples that show that fruitful rational discussions may continue between devotees of a newly established dominant theory and unconvinced sceptics. Such is Galileo's *Two Principal Systems*. Such are some of Einstein's 'popular' writings, or the important criticism of Einstein's principle of covariance voiced by E. Kretschmann (1917), or the criticism of Einstein's General Theory recently voiced by Dicke. And such are Einstein's famous discussions with Bohr. It would be quite incorrect to say that the latter were not fruitful, for not only did Bohr claim that they much improved his understanding of quantum mechanics, but they also led to the famous paper of Einstein, Podolsky, and Rosen which has produced a whole literature of considerable significance, and may yet lead to more.[23] No paper which is discussed by recognized experts for thirty-five years can be denied its scientific status and significance, but this paper was, surely, criticizing (from the outside) the whole framework which had been established by the revolution of 1925–6. Opposition to this framework – the Copenhagen framework – is continued by a minority to which for example de Broglie, Bohm, Landé, and Vigier belong, apart from those names mentioned in the preceding note.[24]

Thus discussions may go on all the time. And although there are always attempts to transform the society of scientists into a closed society, these attempts have not succeeded. In my opinion they would be fatal for science.

The proponents of the myth of the framework distinguish sharply between rational periods of science conducted within a framework (which can be described as periods of closed or authoritarian science) and periods of crisis and revolution (which can be described as the almost irrational leap – comparable to a religious conversion – from one framework to another).

No doubt there are such irrational leaps, such conversions, as described. No doubt there are scientists (normal scientists, presumably) who just follow the lead of others, or give way to social pressure and accept a new theory as a new faith because the experts, the authorities, have accepted it. I admit, regretfully, that there are fashions in science, and that there is also social pressure.

I even admit that the day may come when the social community of scientists will consist mainly or exclusively of scientists who uncritically accept a ruling dogma. They will normally be swayed by fashions. And they will accept a theory because it is the latest cry, and because they fear to be regarded as laggards.

I assert, however, that this will be the end of science as we know it – the end of the tradition created by Thales and Anaximander and rediscovered by Galileo. As long as science is the search for truth it will be the rational critical discussion between competing theories, and the rational critical discussion of the revolutionary theory. This discussion decides whether or not the new theory is to be regarded as better than the old theory: that is, whether or not it is to be regarded as a step towards the truth.

XIV

Almost forty years ago I stressed that even observations, and reports of observations, are under the sway of theories or, if you like, under the sway of a framework. Indeed, there is no such thing as an uninterpreted observation, an observation which is not theory-impregnated. In fact, our very eyes and ears are the result of evolutionary adaptations – that is, of the method of trial and error corresponding to the method of conjectures and refutations. Both methods are adjustments to environmental regularities. A simple example will show that ordinary visual experiences have a pre-Parmenidian absolute sense of up and down built into them – a sense which is no doubt genetically based. The example is this. A square standing on one of its sides looks to all of us a different figure from a square standing on one of its corners. There is a real *Gestalt* shift in moving from one figure to the other.

But I assert that the fact that observations are theory-impregnated does not lead to incommensurability between either observations or theories. For the old observations can be consciously reinterpreted: we can learn that the two squares are different positions of the same square. This is made even easier just because of the genetically based interpretations: no doubt we understand each other so well partly because we share so many physiological mechanisms which are built into our genetic system.

Yet I assert that it is possible for us to transcend even our genetically based physiology. This we do by the critical method. We can understand even a bit of the language of the bees. Admittedly, this understanding is conjectural and rudimentary. But almost all understanding is conjectural, and the deciphering of a new language is always rudimentary to start with.

It is the method of science, the method of critical discussion, which makes it possible for us to transcend not only our culturally acquired but even our inborn frameworks. This method has made us transcend not only our senses but also our partly innate tendency to regard the world as a universe of identifiable things and their properties. Ever since Heraclitus there have been revolutionaries who have told us that the world consists of processes, and that things are things only in appearance: in reality they are processes. This shows how critical thought can challenge and transcend a framework even if it is rooted not only in our conventional language but in our genetics – in what may be called human nature itself. Yet even this revolution does not produce a theory incommensurable with its predecessor: the very task of the revolution was to explain the old category of thing-hood by a theory of greater depth.

XV

I may perhaps also mention that there is a very special form of the myth of the framework which is particularly widespread. It is the view that, before discussion, we should agree on our vocabulary – perhaps by 'defining our terms'.

I have criticized this view on various occasions and I do not have space to do so again.[25] I only wish to make clear that there are the strongest possible reasons against this view. All definitions, so-called 'operational definitions' included, can only shift the problem of the meaning of the term in question to the defining terms. Thus the demand for definitions leads to an infinite regress unless we admit so-called 'primitive' terms, that is, *undefined* terms. But these are as a rule no less problematic than most of the defined terms.

XVI

In the last section of this paper, I will briefly discuss the myth of the framework from a logical point of view: I will attempt something like a logical diagnosis of the malaise.

The myth of the framework is clearly the same as the doctrine that one cannot rationally discuss anything that is *fundamental*, or that a rational discussion of *principles* is impossible.

This doctrine is, logically, an outcome of the mistaken view that all rational discussion must start from some *principles* or, as they are often called, *axioms*, which in their turn must be accepted dogmatically if we wish to avoid an infinite regress – a regress due to the alleged fact that when rationally discussing the validity of our principles or axioms we must again appeal to principles or axioms.

Usually those who have seen this situation either insist dogmatically upon the truth of a framework of principles or axioms, or they become relativists: they say that there are different frameworks and that there is no rational discussion between them, and thus no rational choice.

But all this is mistaken. For behind it there is the tacit assumption that a rational discussion must have the character of a justification, or of a proof, or of a demonstration, or of a logical derivation from admitted premises. But the kind of discussion which is going on in the natural sciences might have taught our philosophers that there is also another kind of rational discussion: a critical discussion which does not seek to prove or to justify or to establish a theory, least of all by deriving it from some higher premises, but which tries to test the theory under discussion by finding out whether its *logical consequences* are all acceptable, or whether it has, perhaps, some undesirable consequences.

We thus can logically distinguish between *a mistaken method of criticizing* and *a correct method of criticizing*. The *mistaken method* starts from the question: How can we establish or justify our thesis or our theory? It thereby leads either to dogmatism, or to an infinite regress, or to the relativistic doctrine of rationally incommensurable frameworks. By contrast, the *correct method* of critical discussion starts from the question: What are the *consequences* of our thesis or our theory? Are they all acceptable to us?

It thus consists in comparing the consequences of different theories (or, if you like, of different frameworks) and tries to find out which of the competing theories or frameworks has consequences that seem preferable to us. It is thus conscious of the fallibility of all our methods, although it tries to replace all our theories by better ones. This is, admittedly, a difficult task, but by no means an impossible one.

Of course, a proponent of the myth of the framework might criticize this idea. He might say, for example, that what I have called the correct method of criticism in no way allows us to get out of our framework – for, he might insist, the 'consequences that seem preferable to us' will themselves be *part* of our framework: that we have here a model for mere self-justification, rather than the critical transcendence of a framework.

But I think that this criticism is mistaken. While we *may* interpret our views in this way, we do not *have* to do so. We can choose to pursue an aim or goal – such as the aim of understanding better the universe in which we live, and ourselves as part of it – which is autonomous of the particular theories or frameworks that we construct to try to meet this aim. And we can choose to set ourselves standards of explanation, and methodological rules, which will help us to achieve our goal and which it is *not* easy for any theory or framework to satisfy. Of course, we may choose not to do this: we may decide to make our ideas self-reinforcing. We may set ourselves no task other than one we know our present ideas can fulfil. We certainly can choose to do this. But if we do choose to do this, not only will we be turning our backs on the possibility of learning that we are wrong, we will also be turning our backs upon that tradition of critical thought (stemming from the Greeks and from culture clash) which has made us what we are, and which offers us the hope of further self-emancipation through knowledge.

To sum up, frameworks, like languages, may be barriers. They may even be prisons. But a strange conceptual framework, just like a foreign language, is no absolute barrier: we can break into it, just as we can break out of our own framework, our own prison. And just as breaking through a language barrier is difficult but very much worth our while, and likely to repay our efforts not only by widening our intellectual horizon but also by offering us much enjoyment, so it is with breaking through the barrier of a framework. A breakthrough of this kind is a discovery for us. It has often led to a breakthrough in science, and it may do so again.

NOTES

1 Herodotus, III, 38.
2 I discuss the distinction between nature and convention in my *Open Society*, chapter 5, where I refer to Pindar, Herodotus, Protagoras, Antiphon, Archelaus, and especially to Plato's *Laws* (cp. notes 3, 7, 10, 11, and 28 to chapter 5 and text). Although I mention (p. 60) the significance of 'the realization that taboos are different in various tribes', and although I (just) mention Xenophanes (note 7) and his profession as a 'wandering bard' (note 9 to chapter 10), I did not then fully realize the part played by culture clash in the evolution of critical thought, as witnessed by the contribution made by Xenophanes, Heraclitus, and Parmenides (see especially note 11 to the *Open Society*, chapter 5) to the problem of nature or reality or truth *versus* convention or opinion. See also my *Conjectures and Refutations, passim*.
3 For further discussion, see my *Open Society* and my *Conjectures and Refutations* (introduction and chapters 4 and 5).

4 Cp. my *Conjectures and Refutations*, pp. 152f. The first two lines of my text are fragment B 16 and the next four fragments B 15. The remaining three fragments are B 18, 35, and 34 (according to H. Diels and W. Kranz, *Fragmente der Vorsokratiker*, 5th edition, Weidmann, Berlin, 1934; see also H. Fränkel, section 4 of *The Pre-Socratics*, edited by A.P.D. Mourelatos, Doubleday Anchor, New York, 1974). The translations are mine. Note, in the last quoted two lines, the contrast between the one final truth and the many guesses, or opinions, or conjectures.

5 Parmenides used Xenophanes' terminology. See my *Conjectures and Refutations*, for example pp. 11, 17, 145, 400, and 410. See also my *Open Society*, note 56 (section 8) to chapter 10.

6 See Parmenides' remark (in fragment B 6) on the muddled horde of erring mortals, always in two minds about things, in contrast to the one 'well-rounded truth'. Cp. my *Conjectures and Refutations*, pp. 11, 164f.

7 *Theogony*, 720–5.

8 *Iliad*, VIII, 13–16; cp. *Aeneid*, VI, 577.

9 See my *Conjectures and Refutations*, pp. 126ff., 138f., 150f., 413.

10 *Theogony*, 720–5.

11 The discovery, it appears, is due to Parmenides. See fragments B14–15:

> Bright'ning the night she glides round the earth with a light which is borrowed;
> Always she wistfully looks round for the rays of the sun.

12 Conventionality is not quite the same as mere arbitrariness. For there may be better or worse conventions. See my *Open Society*, chapter 5, esp. pp. 64f.

13 Hegel distinguishes, of course, between 'appearance' and 'reality'. (Wallace translates '*Wirklichkeit*' not by 'reality' but by 'actuality' – see, for example, his *Logic of Hegel*, Oxford, 1874, p. 7.) God is 'most real'. He is 'that which alone is truly real', what exists accidentally is only 'appearance'. Hegel writes: 'Who would not be clever enough to see that much in his environment is not what it ought to be?' And he says that philosophy 'has to do only with the Idea, which is not so powerless that it only determines what ought to be and not what really is.' (The quotations are from G.W.F. Hegel, *Encyclopädie der philosophischen Wissenschaften im Grundrisse; Die Logik*, Einleitung, § 6. See, for example, the Henning edition, Duncker and Humblot, Berlin, 1840, pp. 9–11. Cp. W. Wallace, *The Logic of Hegel*, pp. 7–9.) This is, of course, sufficient to muddle what is and what ought to be, and thus to defend practically any view (and perhaps its opposite).

14 See Alfred Tarski, *Logic, Semantics, Metamathematics*, translated by J.H. Woodger, Oxford University Press, London, 1956. I have expounded it in various places. See, for example, my *Conjectures and Refutations*, pp. 223–5.

15 See Benjamin Lee Whorf, *Language, Thought, and Reality*, edited by John B. Carroll, MIT Press, Cambridge, Mass., 1956.

16 See W.V.O. Quine, *Word and Object*, MIT Press, Cambridge, Mass., 1960, and *Ontological Relativity and Other Essays*, Columbia University Press, New York, 1969.

17 I completely agree with Quine's criticism of the museum theory (or zoo theory) of the meaning of words – the theory that the world is *one* museum with labelled showcases whose contents are the unambiguous referents of the words or labels. But there may be different museums. And the relevance of the contents of the showcases may depend on history – for example, on the problems which have been thrown up. But I am just as critical of any observationalist or behaviourist theory of the meaning of words. Translation, I conjecture, is a matter of conjecture and refutation – of conjecturing the other fellow's *problems* and his background knowledge. This, I suggest, is the way a first language or a second language is learned. Observation of behaviour may pose problems, and may help by way of refutation. (See my *Unended Quest*, section 7.)

18 Cp. p. 232 of T.S. Kuhn, 'Reflections on my Critics', in *Criticism and the Growth of Knowledge*, edited by I. Lakatos and A. Musgrave, Cambridge University Press, London, 1970, pp. 231–78.

19 When writing this section, I originally had in mind Thomas Kuhn and his book *The Structure of Scientific Revolutions*, Chicago University Press, Chicago, 1962, 1970. See also my contribution 'Normal Science and its Dangers' to *Criticism and the Growth of Knowledge*, pp. 51–8. However, as Kuhn points out, this interpretation was based on a misunderstanding of his views (see his 'Reflections on my Critics', in *Criticism and the Growth of Knowledge*, pp. 231–78, and his 'Postscript 1969' to the 2nd edition of *The Structure of Scientific Revolutions*), and I am very ready to accept his correction. Nevertheless, I regard the view here discussed as influential.

20 For a criticism of Karl Mannheim, see chapters 23 and 24 of my *Open Society*.

21 Few people seem to realize that by his equation $E=mc^2$, Einstein resurrected the fluidum theory of heat (caloric) for which the question whether heat has any weight was regarded as crucial. According to Einstein's theory, heat *has* weight – only it weighs very little.

22 Cp. the discussion between Planck and Mach, especially Planck's paper 'Zur Machschen Theorie der physikalischen Erkenntnis', *Physikalische Zeitschrift*, *II*, 1910, pp. 1186–90.

23 See, for example, J.S. Bell, 'On the Einstein Podolsky Rosen Paradox', *Physics*, *1*, 1964, pp. 195-200, and 'On the Problem of Hidden Variables in Quantum Mechanics', *Reviews of Modern Physics*, *38*, 1966, pp. 447–52. See also John F. Clauser, Michael A. Horne, Abner Shimony, and Richard A. Holt, 'Proposed Experiment to Test Local Hidden Variable Theories', *Physical Review Letters*, 13 October 1969. An extension or strengthening of the EPR paradox, described in my *Logic of Scientific Discovery*, pp. 446–8, seems to me to involve a decisive refutation of the Copenhagen interpretation, since the two simultaneous measurements together would allow simultaneous 'reductions' of the two wave packets which cannot be carried out within the theory. See also the paper by James Park and Henry Margenau, 'Simultaneous Measurability in Quantum Theory', *International Journal of Theoretical Physics*, *1*, 1968, pp. 211–83.

24 See my 'Quantum Mechanics without "The Observer"', in *Studies in the Foundations, Methodology and Philosophy of Science*, volume II: *Quantum Theory and Reality*, edited by Mario Bunge, Springer-Verlag, New York, 1967.

25 See my *Open Society*, chapter 11, section II, or my 'Quantum Mechanics without "The Observer"', especially pp. 11–15, or my *Conjectures and Refutations*, pp. 19, 28 (section 9), 279, and 402.

3

REASON OR
REVOLUTION?

The trouble with a total revolution ...
Is that it brings the same class up on top.
Executives of skilful execution
Will therefore plan to go halfway and stop.

Robert Frost

The following critical considerations are reactions to the book, *Der Positivismusstreit in der deutschen Soziologie*,[1] which was published in 1969, and for which I unwittingly provided the original incentive.

I

I will begin by telling some of the history of the book and of its misleading title. In 1960 I was invited to open a discussion on 'The Logic of the Social Sciences' at a congress of German sociologists in Tübingen. I accepted – and I was told that my opening address would be followed by a reply from Professor Theodor W. Adorno

This paper came into being as a result of a suggestion by Professor Raymond Aron. My paper 'The Logic of the Social Sciences' was first published in Germany as the third paper of a collection misnamed *Der Positivismusstreit in der deutschen Soziologie* (see note 1, below), in a manner which left it unexplained that it was the paper which had unwittingly sparked off this '*Positivismusstreit*'. In 1970 I wrote a letter to the *Times Literary Supplement* ('Dialectical Methodology', *TLS 69*, 26 March 1970, pp. 388–9) in criticism of a review of the *Positivismusstreit* volume that had appeared there. Professor Aron suggested that I expand this letter, and explain my objections to the volume more fully. This I did in the present paper, which was first published in *Archives européennes de sociologie, 11*, 1970, pp. 252–62, and which is also appended to the English translation of the *Positivismusstreit* volume. (See Theodor W. Adorno, *et al.*, eds, *The Positivist Dispute in German Sociology*, translated by Glyn Adey and David Frisby, Harper & Row, 1976.) The motto is from Robert Frost, 'A Semi-Revolution', in *A Witness Tree*.

65

of Frankfurt. It was suggested to me by the organizers that, in order to make a fruitful discussion possible, I should formulate my views in a number of definite theses. This I did: my opening address to that discussion, delivered in 1961, consisted of twenty-seven sharply formulated theses, plus a programmatic formulation of the task of the theoretical social sciences. Of course, I formulated these theses so as to make it difficult for any Hegelian or Marxist (such as Adorno) to accept them. And I supported them as well as I could by arguments. Owing to the limited time available, I confined myself to fundamentals, and I tried to avoid repeating what I had said elsewhere.

Adorno's reply was read with great force, but he hardly took up my challenge – that is, my twenty-seven theses. In the ensuing debate Professor Ralf Dahrendorf expressed his grave disappointment. He said that it had been the intention of the organizers to bring into the open some of the glaring differences – apparently he included political and ideological differences – between my approach to the social sciences and Adorno's. But the impression created by my address and Adorno's reply was, he said, one of sweet agreement – a fact which left him flabbergasted ('*als seien Herr Popper und Herr Adorno sich in verblüffender Weise einig*'). I was and I still am very sorry about this. But having been invited to speak about 'The Logic of the Social Sciences' I did not go out of my way to attack Adorno and the 'dialectical' school of Frankfurt (Adorno, Horkheimer, Habermas, *et al.*), which I never regarded as important, unless perhaps from a political point of view. I was not aware of the organizers' intention, and in 1960 I was not even aware of the political influence of this school. Although today I would not hesitate to describe this influence by such terms as 'irrationalist' and 'intelligence-destroying', I could never take their methodology (whatever that may mean) seriously from either an intellectual or a scholarly point of view. Knowing now a little more, I think that Dahrendorf was right in being disappointed: I ought to have attacked them using arguments I had previously published in my *Open Society* and *The Poverty of Historicism* and in 'What is Dialectic?',[2] even though I do not think that these arguments fall under the heading of 'The Logic of the Social Sciences' – for terms do not matter. My only comfort is that the responsibility for avoiding a fight rests squarely on the second speaker.

However this may be, Dahrendorf's criticism stimulated a paper (almost twice as long as my original address) by Professor Jürgen

Habermas, another member of the Frankfurt School. It was in this paper, I think, that the term 'positivism' first turned up in this particular discussion: I was criticized as a *positivist*. This is an old misunderstanding created and perpetuated by people who know of my work only at second-hand. Owing to the tolerant attitude adopted by some members of the Vienna Circle, my book, *Logik der Forschung*, in which I criticized this positivist Circle from a realist and anti-positivist point of view, was published in a series of books edited by Moritz Schlick and Philipp Frank, two leading members of the Circle.[3] And those who judge books by their covers (or by their editors) created the myth that I had been a member of the Vienna Circle and a positivist. Nobody who has read that book (or any other book of mine) would agree – unless indeed he believed in the myth to start with, in which case he may of course find evidence to support his belief.

In my defence Professor Hans Albert (not a positivist either) wrote a spirited reply to Habermas' attack. The latter answered, and was rebutted a second time by Albert. This exchange was mainly concerned with the general character and tenability of my views. Thus there was little mention – and no serious criticism – of my opening address of 1961, and of its twenty-seven theses.

It was, I think, in 1964 that a German publisher asked me whether I would agree to have my address published in book form together with Adorno's reply and the debate between Habermas and Albert. I agreed.

But as now published (in 1969, in German), the book consists of two quite new introductions by Adorno (94 pages), followed by my address of 1961 (20 pages) with Adorno's original reply (18 pages), Dahrendorf's complaint (9 pages), the debate between Habermas and Albert (150 pages), a new contribution by Harold Pilot (28 pages), and a 'Short Surprised Postscript to a Long Introduction' by Albert (5 pages). In this, Albert mentions briefly that the affair started with a discussion between Adorno and myself in 1961, and he says quite rightly that a reader of the book would hardly realize what it was all about. This is the only allusion in the book to the story behind it. There is no answer to the question of how the book got a title which quite wrongly indicates that the opinions of some 'positivists' are discussed in it. Even Albert's postscript does not answer the question.

What is the result? My twenty-seven theses, intended to start a discussion (and so they did, after all), are nowhere seriously taken up in this longish book – not a single one of them, although one or another passage from my address is mentioned here or there, usually out of context, to illustrate my 'positivism'. Moreover, my address is buried in the middle of the book, unconnected with the beginning and the end. No reader can see, and no reviewer can understand, why my address (which I cannot but regard as quite unsatisfactory in its present setting) is included in the book – or that it is the unadmitted theme of the whole book. Thus no reader would suspect, and no reviewer did suspect, what I suspect as being the truth of the matter. It is that my opponents literally did not know how to criticize rationally my twenty-seven theses. All they could do was to label me 'positivist' (thereby unwittingly giving a highly misleading name to a debate in which not a single 'positivist' was involved). And having done so, they drowned my short paper, and the original issue of the debate, in an ocean of words – which I found only partially comprehensible.

As it now stands, the main issue of the book has become Adorno's and Habermas' accusation that a 'positivist' like Popper is bound by his methodology to defend the political *status quo*. It is an accusation which I myself raised in my *Open Society* against Hegel, whose identity philosophy (what is real is reasonable) I described as 'moral and legal positivism'. In my address I had said nothing about this issue, and I had no opportunity to reply. But I have often combated this form of 'positivism' along with other forms. And it is a fact that my *social theory* (which favours gradual and piecemeal reform, reform controlled by a critical comparison between expected and achieved results) contrasts with my *theory of method*, which happens to be a theory of scientific and intellectual revolution.

II

This fact and my attitude towards revolution can be easily explained. We may start from Darwinian evolution. Organisms evolve by trial and error, and their erroneous trials – their erroneous mutations – are eliminated, as a rule, by the elimination of the organism which is the 'carrier' of the error. It is part of my epistemology that, in man, through the evolution of a descriptive and argumentative language, all this has changed radically. Man has achieved the possibility of being *critical of his own tentative trials, of his own theories*. These theories are no longer incorporated in his organism or in his genetic system. They may be formulated in books or in journals. And they can be critically discussed, and shown to be erroneous, without killing any authors or burning books – without destroying the 'carriers'.

In this way we arrive at a fundamental new possibility: *our trials, our tentative hypotheses, may be critically eliminated by rational discussion, without eliminating ourselves*. This indeed is the purpose of rational critical discussion.

The 'carrier' of a hypothesis has an important function in these discussions: he has to defend the hypothesis against erroneous criticism, and he may perhaps try to modify it if in its original form it cannot be successfully defended.

If the method of rational critical discussion should establish itself, then this should make the use of violence obsolete. *For critical reason is the only alternative to violence so far discovered.*

It is the obvious duty of all intellectuals to work for *this* revolution – for the replacement of the eliminative function of violence by the eliminative function of rational criticism. But to work for this end, one has to train oneself constantly to write and speak in clear and simple language. Every thought should be formulated as clearly and simply as possible. This can only be achieved by hard work.

III

I have been for many years a critic of the so-called 'sociology of knowledge'. Not that I thought that everything that Mannheim (and Scheler) said was mistaken. On the contrary, much of it was only too trivially true. What I combated was Mannheim's belief that there was an essential difference with respect to objectivity between the social scientist and the natural scientist, or between the study of society and the study of nature. The thesis I combated was that it was easy to be objective in the natural sciences, while objectivity in the social sciences could be achieved, if at all, only by very select intellects: by the 'freely poised intelligence' which is only 'loosely anchored in social traditions'.[4]

As against this I stressed that the objectivity of natural and social science is not based on an impartial state of mind in the scientists, but merely on the fact of the public and competitive character of the scientific enterprise and thus on certain social aspects of it. This is why I wrote: '*What the "sociology of knowledge" overlooks is just the sociology of knowledge* – the social or public character of science.'[5] Objectivity is based, in brief, upon *mutual rational criticism*, upon the critical approach, the critical tradition.[6]

Thus natural scientists are not more objectively minded than social scientists. Nor are they more critical. If there is more objectivity in the natural sciences, this is because there is a better tradition and higher standards of clarity and of rational criticism.

In Germany, many social scientists are brought up as Hegelians, and this is a tradition destructive of intelligence and critical thought. It is one of the points where I agree with Karl Marx who wrote: 'In its mystifying form dialectic became the ruling German fashion'.[7] It is the German fashion still.

IV

The sociological explanation of this fact is simple. We all get our values, or most of them, from our social environment: often merely by imitation (simply by taking them over from others), sometimes by a revolutionary reaction to accepted values, and at other times – though this may be rare – by a critical examination of these values and of possible alternatives. However this may be, the social and intellectual climate, the tradition in which one is brought up, is often decisive for the moral and other standards and values one adopts. All this is rather obvious. A very special case, but all-important for our purpose, is that of intellectual values.

Many years ago I used to warn my students against the widespread idea that one goes to college in order to learn how to talk and write 'impressively' and incomprehensibly. At the time many students came to college with this ridiculous aim in mind, especially in Germany. And most of those students who, during their university studies, enter into an intellectual climate which accepts this kind of valuation – coming, perhaps, under the influence of teachers who in their turn had been reared in a similar climate – are lost. They unconsciously learn and accept that highly obscure and difficult language is the intellectual value *par excellence*. There is little hope that they will even understand that they are mistaken, or that they will ever realize that there are other standards and values – values such as truth, the search for truth, the approximation to truth through the critical elimination of error, and clarity. Nor will they find out that the standard of 'impressive' obscurity actually clashes with the standards of truth and rational criticism. For these latter values depend on clarity. One cannot tell truth from falsity, one cannot tell an adequate answer to a problem from an irrelevant one, one cannot tell good ideas from trite ones, and one cannot evaluate ideas critically – unless they are presented with sufficient clarity. But to those brought up in the implicit admiration of brilliance and 'impressive' opaqueness, all this (and all I have said here) would be *at best*, 'impressive' talk: they do not know any other values.

Thus arose the cult of incomprehensibility, of 'impressive' and high-sounding language. This was intensified by the (for laymen) impenetrable and impressive formalism of mathematics. I suggest that in some of the more ambitious social sciences and philosophies, especially in Germany, the traditional game, which has largely become the unconscious and unquestioned standard, is to state the utmost trivialities in high-sounding language.

If those brought up on this kind of nourishment are presented with a book which is written simply and contains something unexpected, controversial, or new, they usually find it difficult or impossible to understand. For it does not conform to their idea of 'understanding', which for them entails agreement. That there may be important ideas worth understanding with which one cannot at once agree or disagree is to them unfathomable.

V

There is here, at first sight, a difference between the social sciences and the natural sciences: in the so-called social sciences and in philosophy, the degeneration into impressive but more or less empty verbalism has gone further than in the natural sciences. Yet the danger is getting acute everywhere. Even among mathematicians a tendency to impress people may sometimes be discerned, although the incitement to do so is least here. For it is partly the wish to ape the mathematicians and the mathematical physicists in technicality and in difficulty that inspires the use of verbiage in other sciences.

Yet lack of critical creativeness – that is, of inventiveness paired with critical acumen – can be found everywhere. And everywhere this leads to the phenomenon of young scientists eager to pick up the latest fashion and the latest jargon. These 'normal' scientists[8] want a framework, a routine, a common, and exclusive language of their trade. But it is the non-normal scientist, the daring scientist, the critical scientist, who breaks through the barrier of normality, who opens the windows and lets in fresh air, who does not think about the impression he makes, but tries to be well understood.

The growth of normal science, which is linked to the growth of Big Science, is likely to prevent, or even to destroy, the growth of knowledge, the growth of great science.

The situation is tragic if not desperate. And the present trend in the so-called empirical investigations into the sociology of the natural sciences is likely to contribute to the decay of science. Superimposed upon this danger is another danger, created by Big Science: its urgent need for scientific technicians. More and more Ph.D. candidates receive merely technical training in certain techniques of measurement. They are not initiated into the scientific tradition, the critical tradition of questioning, of being tempted and guided by great and apparently insoluble riddles rather than by the solubility of little puzzles. True, these technicians, these specialists, are usually aware of their limitations. They call themselves 'specialists' and reject any claim to authority outside their specialities. Yet they do so proudly, and proclaim that specialization is a necessity. But this means flying in the face of the facts, which show that great advances still come from those with a wide range of interests.

If the many, the specialists, gain the day, it will be the end of science as we know it – of great science. It will be a spiritual catastrophe comparable in its consequences to nuclear armament.

VI

I now come to my main point. It is this. Some of the famous leaders of German sociology who do their intellectual best, and do it with the best conscience in the world, are nevertheless, I believe, simply talking trivialities in high-sounding language, as they were taught. They teach this to their students, who are dissatisfied, yet do the same. The genuine and general feeling of dissatisfaction, manifest in their hostility to the society in which they live, is a reflection of their unconscious dissatisfaction with the sterility of their own activities.

I will give a brief example from the writings of Professor Adorno. The example is a select one – selected, indeed, by Professor Habermas, who begins his first contribution to *Der Positivismusstreit* by quoting it. On the left I give the original German text, in the centre this text as translated in the present volume, and on the right a paraphrase into simple English of what seems to be being asserted.[9]

Die gesellschaftliche Totalität führt kein Eigenleben oberhalb des von ihr Zusammengefassten, aus dem sie selbst besteht.

Societal totality does not lead a life of its own over and above that which it unites and of which it, in its turn comprises.

Society consists of social relationships.

Sie produziert und reproduziert sich durch ihre einzelnen Momente hindurch. ...

It produces and reproduces itself through its individual moments. ...

The various social relationships somehow produce society. ...

So wenig aber jenes Ganze vom Leben, von der Kooperation und dem Antagonismus seiner Elemente abzusondern ist,

This totality can no more be detached from life, from the cooperation and the antagonism of its elements,

Among these relations are cooperation and antagonism; and since society consists of these relations, it is impossible to separate it from them.

so wenig kann irgendein Element auch bloss in seinem Funktionieren verstanden werden ohne Einsicht in des Ganze, das an der Bewegung des Einzelnen selbst sein Wesen hat.

than can an element be understood merely as it functions without insight into the whole which has its source (*Wesen*, essence) in the motion of the individual entity itself.

The opposite is also true: none of the relations can be understood without the totality of all the others.

System und Einzelheit sind reziprok und nur in ihre Reziprozität zu erkennen.

System and individual entity are reciprocal and can only be apprehended in their reciprocity.

(Repetition of the preceding thought.)

Comment: The theory of the social wholes developed here has been presented and developed, sometimes better and sometimes worse, by countless philosophers and sociologists. I do not assert that it is mistaken. I only assert the complete triviality of its content. Of course Adorno's *presentation* is very far from trivial.

VII

It is for reasons such as these that I find it so difficult to discuss any serious problem with Professor Habermas. I am sure he is perfectly sincere. But I think that he does not know how to put things simply, clearly, and modestly, rather than impressively. Most of what he says seems to me trivial. The rest seems mistaken.

So far as I can understand him, the following is his central complaint about my alleged views. My way of theorizing, Habermas suggests, violates the *principle of the identity of theory and practice* – perhaps because I say that theory should *help* action, that is, should help us to modify our actions. For I say that it is the task of the theoretical sciences to try to anticipate the unintended consequences of our actions. Thus I differentiate between this theoretical task and the action. But Professor Habermas seems to think that only one who is a practical critic of existing society can produce serious theoretical arguments about society, since social knowledge cannot be divorced from fundamental social attitudes. The indebtedness of this view to the sociology of knowledge is obvious and need not be laboured.

My reply is very simple. We should welcome any suggestion as to how our problems might be solved, regardless of the attitude towards society of the man who puts them forward: provided that he has learned to express himself clearly and simply – in a way that can be understood and evaluated – and that he is aware of our fundamental ignorance and responsibilities towards others. But I do not think that the debate about the reform of society should be reserved for those who first put in a claim for recognition as practical revolutionaries, and who see the sole function of the revolutionary intellectual in pointing out as much as possible what is repulsive in our social life (excepting their own social roles).

It may be that revolutionaries have a greater sensitivity to social ills than other people. But obviously, there can be better and worse revolutions (as we all know from history), and the problem is not to do too badly. Most, if not all, revolutions have produced societies very different from those desired by the revolutionaries. *Here is a problem*, and it deserves thought from every serious critic of society. And this should include an effort to put one's ideas into simple, modest language, rather than high-sounding jargon. This is an effort which those fortunate ones who are able to devote themselves to study owe to society.

VIII

A last word about the term 'positivism'. Words do not matter, and I do not really mind if even a thoroughly misleading and mistaken label is applied to me. But the fact is that throughout my life I have combated positivist epistemology, under the name 'positivism'. I do not deny, of course, the possibility of stretching the term 'positivist' until it covers anybody who takes any interest in natural science, so that it can be applied even to opponents of positivism, such as myself. I only contend that such a procedure is neither honest nor apt to clarify matters.

The fact that the label 'positivism' was originally applied to me by sheer blunder can be checked by anybody who is prepared to read my early *Logik der Forschung*.

It is, however, worth mentioning that one of the victims of the two misnomers, 'positivism' and *'Der Positivismusstreit'* is Dr Alfred Schmidt, who describes himself as a 'collaborator of many years' standing' (*langjähriger Mitarbeiter*) of Professors Adorno and Horkheimer. In a letter to a newspaper *Die Zeit*,[10] written to defend Adorno against the suggestion that he misused the term 'positivism' in *Der Positivismusstreit* or on similar occasions, Schmidt characterizes positivism as a tendency of thought in which 'the method of the various single sciences is taken absolutely as the only valid method of knowledge' (*die einzelwissenschaftlichen Verfahren als einzig gültige Erkenntnis verabsolutierende Denken*), and he identifies it, correctly, with an over-emphasis on 'sensually ascertainable facts'. He is clearly unaware of the fact that my alleged positivism, which was used to give the book *Der Positivismusstreit* its name, consisted of a fight against all this, which he describes (fairly correctly) as 'positivism'. I have always fought for the right to operate freely with speculative theories against the narrowness of the 'scientistic' theories of knowledge and, especially, against all forms of sensualistic empiricism.

I have fought against the aping of the natural sciences by the social sciences,[11] and I have fought for the doctrine that positivistic epistemology is inadequate even in its analysis of the natural sciences which, in fact, are not 'careful generalizations from observation', as is usually believed, but are essentially speculative and daring. Moreover, I have taught, for more than thirty-eight years,[12] that all observations are theory-impregnated, and that their main function is to check and refute, rather than to prove, our theories. Finally I have not only stressed the meaningfulness of metaphysical assertions and the fact that I am myself a metaphysical realist, but I have also analysed the important historical role played by metaphysics in the formation of scientific theories. Nobody before Adorno and

Habermas has described such views as 'positivistic', and I can only suppose that these two did not know, originally, that I held such views. (In fact, I suspect that they were no more interested in my views than I am in theirs.)

The suggestion that anybody interested in natural science is to be condemned as a positivist would make positivists not only of Marx and Engels, but also of Lenin – the man who introduced the equation of 'positivism' and 'reaction'.

Terminology does not matter, however. Only it should not be used as an *argument*. And the title of a book ought not to be dishonest – nor should it attempt to prejudge an issue.

On the substantial issue between the Frankfurt School and myself – revolution versus piecemeal reform – I shall not comment here, since I have treated it as well as I could in my *Open Society*. Hans Albert too has said many incisive things on this topic, both in his replies to Habermas in *Der Positivismusstreit* and in his important book *Traktat über kritische Vernunft*.[13]

NOTES

1 H. Maus and F. Fürstenberg, eds, *Der Positivismusstreit in der deutschen Soziologie*, Luchterhand, Berlin, 1969.
2 'What is Dialectic?', *Mind*, XLIX, 1940, pp. 403ff. Reprinted in *Conjectures and Refutations*.
3 The Vienna Circle consisted of men of originality and of the highest intellectual and moral standards. Not all of them were positivists, even if we mean by this term no more than a condemnation of speculative thought, although most of them were. I have always been in favour of criticizable speculative thought and, of course, of its criticism.
4 The quotation is from Mannheim. It is discussed more fully in my *Open Society*, volume ii, p. 215.
5 *The Poverty of Historicism*, p. 155.
6 Cp. my *Conjectures and Refutations*, especially chapter 4.
7 Karl Marx, *Capital*, volume II, 1872, 'Nachwort'. (In some later editions this is described as 'Preface to second edition'. The usual translation is not 'mystifying' but 'mystified'. To me this sounds like a Germanism.)
8 The phenomenon of normal science was discovered, but not criticized, by Thomas Kuhn in *The Structure of Scientific Revolutions*. Kuhn is, I believe, mistaken in thinking that 'normal' science is not only normal *today* but always was so. On the contrary, in the past – until 1939 – science was almost always critical, or 'extraordinary'. There was no scientific *'routine'*.
9 In the original publication of this article in *Archives européennes de sociologie* the three columns contained, respectively, the original German, a paraphrase into simple German of what seemed to be being

3. REASON OR REVOLUTION?

asserted, and a translation of this paraphrase into English.

11 See my *Logic of Scientific Discovery*, new appendix 1.
12 See my *Logic of Scientific Discovery*, new appendix *10.
13 Hans Albert, *Traktat über kritische Vernunft*, J.C.B. Mohr, Tübingen, 1969.

ADDENDUM 1974:
THE FRANKFURT SCHOOL

I first heard of the Frankfurt School in the 1930s but decided then, on the basis of some experimental reading, against conscientiously reading its output.

In 1960, as recounted in my 'Reason or Revolution?', I was asked to open a discussion at a conference in Tübingen, and was told that Adorno would reply to my paper.[1] This led me to another attempt at reading the publications of the Frankfurt School and especially Adorno's books.

Most of Adorno's works may be divided into three groups. First, there are his essays on music, literature, or culture. These I found little to my taste. To me they read like imitations of Karl Kraus, the Viennese writer – bad imitations, because they lacked Kraus' sense of humour. I had known, and heartily disliked, this kind of writing in my days in Vienna. I used to think of it as cultural snobbery, practised by a clique which regarded itself as a cultural *élite*. These essays, incidentally, are characterized by their social irrelevance.

Then there was a second group of books, on epistemology or philosophy. And these seemed just the sort of thing one calls in English 'mumbo-jumbo' (or in German 'Hokuspokus').

Of course, Adorno was a Hegelian as well as a Marxist. And I am opposed to both: to Marxism and especially to Hegelianism.

As to Marx, I have great respect for him as a thinker and as a fighter for a better world, though I disagree with him on many points of decisive importance. I have criticized his theories at considerable length. He is not always particularly easy to understand, but he always tries his best to be understandable. For he has something to say, and he wants people to understand him. But as to Adorno, I can neither agree nor disagree with most of his philosophy. In spite of all efforts to understand his philosophy, it seems to me that all of it, or almost all, is just words. He has nothing whatever to say, and he says it in Hegelian language.

But there is a third group of his writings. The essays which belong

---

In December 1973 I was asked by the BBC if I would agree to be interviewed on the subject of the so-called 'Frankfurt School' for use in a programme on their work to be broadcast in January 1974. I prepared a brief paper (which was not, in fact, broadcast in this form because I was given only five minutes in which to speak) containing a few critical remarks I make in section I of my paper 'Reason or Revolution' about my attitude towards the Frankfurt School.

to this third group are mainly complaints about the times we live in. But some of them are interesting and even moving. They give direct expression to his fears: to his anxiety, as he calls it himself, and to his deep depression. Adorno was a pessimist. After Hitler came to power – an event which, he says, surprised him as a politician – he despaired of mankind, and he surrendered his belief in the Marxist gospel of salvation. It is a voice of utmost despair which sounds from these essays – a tragic and pitiful voice.

But so far as Adorno's pessimism is philosophical, its philosophical content is nil. Adorno is consciously opposed to clarity. Somewhere he even mentions with approval that the German philosopher Max Scheler asked for 'more darkness' (*mehr Dunkel*), alluding to the last words of Goethe, who asked for 'more light' (*mehr Licht*).

It is difficult to understand how a Marxist like Adorno could support a demand for more darkness. Marx, certainly, was for enlightenment. But Adorno has published, together with Horkheimer, a book under the title *Dialectic of Enlightenment*[2] in which they try to show that the very idea of enlightenment leads, by its inner contradictions, into darkness – the darkness which we are allegedly in now. This is, of course, a Hegelian idea. Nevertheless, it remains a puzzle how a socialist, or a Marxist, or a humanist, like Adorno, can revert to such Romantic views, and prefer the maxim 'more darkness' to 'more light'. Adorno acted on his maxim by publishing intentionally obscure and even oracular writings. It can only be explained by the nineteenth-century tradition of German philosophy, and by the 'Rise of Oracular Philosophy', as I call it in my *Open Society* – the rise of the school of the so-called German Idealists. Marx himself was brought up in this tradition, but he reacted forcefully against it, and in *Capital* he made a remark about it, and about Dialectic, which I always admired. Marx said in *Capital*: 'In its mystifying form the Dialectic became the ruling fashion in Germany.'[3] Dialectic is still the ruling fashion in Germany. And it is still 'in its mystifying form'.

But I would like also to say a few words about Horkheimer. Compared with Adorno, his writings are lucidity itself. But Horkheimer's so-called 'Critical Theory' is empty – devoid of content. This is more or less admitted by the editor of Horkheimer's *Kritische Theorie*, when he says: 'To cast Horkheimer's conception into the form of understandable ('*eingängige*') propositions is ... almost impossible'.[4] There remains only a vague and unoriginal

Marxian historicism: Horkheimer does not say anything tenable that has not been said better before. His views may be said to be objectively uninteresting, including those with which I can agree.

For I have found in Horkheimer some propositions with which I *can* agree. I can even agree with Horkheimer's formulation of his ultimate aims. In the second volume of his book *Kritische Theorie* he says, after rejecting Utopianism: 'Nevertheless, the idea of a future society as a community of free men ... has a content to which we ought to remain loyal through all [historical] change.'[5] I certainly agree with this idea, the idea of a society of free men (and also with the idea of loyalty to it). It is an idea that inspired the American and the French revolutions. Unfortunately, Horkheimer has nothing of the slightest interest to say about the problem of how to get nearer to this ideal aim.

In fact, Horkheimer rejects, without argument and in defiance of historical facts, the possibility of reforming our so-called 'social system'. This amounts to saying: Let the present generation suffer and perish -- for all we can do is to expose the ugliness of the world we live in, and to heap insults on our oppressors, the 'bourgeoisie'. *This is the total content of the so-called Critical Theory of the Frankfurt School.*

Marx's own condemnation of our society makes sense. For Marx's theory contains the promise of a better future. But the theory becomes vacuous and irresponsible if this promise is withdrawn, as it is by Adorno and Horkheimer. This is why Adorno found that life is not worth living. For life is really worth living only if we can work for a better world *now*, and for the immediate future.

It is a crime to exaggerate the ugliness and the baseness of the world: it is ugly, but it is also very beautiful; inhuman, and also very human. And it is threatened by great dangers. The greatest is world war. Almost as great is the population explosion. But there is much that is good in this world. For there is much good will. And there are millions of people alive today who would gladly risk their lives if they thought that they could thereby bring about a better world.

We can do much *now* to relieve suffering and, most important, to increase individual human freedom. We *must not* wait for a goddess of history or for a goddess of revolution to introduce better conditions into human affairs. History, and also a revolution, may easily fail us. It did fail the Frankfurt School, and it caused Adorno to despair. We must produce and critically try out ideas about what can and should be done now – and do it now.

To sum up with a phrase of Raymond Aron, I regard the writings of the Frankfurt School as 'opium of the intellectuals'.[6]

NOTES

1 See *Archives européennes de sociologie 11*, 1970, pp. 252–62; or the revised version of this paper in *The Positivist Dispute in German Sociology*.
2 Max Horkheimer and Theodor W. Adorno, *Dialectic of Enlightenment*, Herder & Herder, New York, 1972.
3 Karl Marx, *Capital*, volume II, 1872, 'Nachwort'.
4 Max Horkheimer, *Kritische Theorie*, edited by A. Schmidt, S. Fisher, Frankfurt, 1968, volume II, pp. 340f.
5 Horkheimer, *Kritische Theorie*, p. 166.
6 Raymond Aron, *L'Opium des intellectuels*, Calmann-Lévy, Paris, 1955.

4

SCIENCE:
PROBLEMS, AIMS,
RESPONSIBILITIES

I

The intellectual history of man has its depressing as well as its exhilarating aspects. For one may well look upon it as a history of prejudice and dogma, tenaciously held, and often combined with intolerance and fanaticism. One may even describe it as a history of spells of religious or quasi-religious frenzy. It should be remembered, in this context, that most of our great destructive wars have been religious or ideological wars – with the notable exception, perhaps, of the wars of Genghis Khan, who seems to have been a model of religious toleration.

Yet even the sad and depressing picture of religious wars has its brighter side. It is an encouraging fact that countless men, from ancient to modern times, have been ready to live and to die for their convictions, for ideas – ideas which they believed to be true.

Man, we may say, appears to be not so much a rational animal as an ideological animal.

The history of science, even of modern science since the Renaissance, and especially since Francis Bacon, may be taken as an illustration. The movement inaugurated by Bacon was a religious or semi-religious movement, and Bacon was the prophet of the secularized religion of science. He replaced the name 'God' by the name 'Nature', but he left almost everything else unchanged. Theology, the science of God, was replaced by the science of Nature.

Revised version of an address to the Plenary Session of the 47th Annual Meeting of the Federation of American Societies for Experimental Biology, Atlantic City, NJ, 17 April 1963, first published in *Federation Proceedings*, 22, 1963, pp. 961–72.

The laws of God were replaced by the laws of Nature. God's power was replaced by the forces of Nature. And at a later date, God's design and God's judgments were replaced by natural selection. Theological determinism was replaced by scientific determinism, and the book of fate by the predictability of Nature. In short, God's omnipotence and omniscience were replaced by the omnipotence of nature and by the virtual omniscience of natural science.

It was also in this period that the phrase '*deus sive natura*' – which may perhaps be translated as 'God, or what is the same, nature' – was almost casually used by the physicist and philosopher Spinoza.

According to Bacon, nature, like God, was present in all things, from the greatest to the least. And it was the aim or the task of the new science of nature to determine the nature of all things, or, as he sometimes said, the essence of all things. This was possible because the Book of Nature was an open book. All that was needed was to approach the goddess Nature with a pure mind, free of prejudices, and she would readily yield her secrets. Give me a couple of years free from other duties, Bacon somewhat unguardedly exclaimed in a moment of enthusiasm, and I shall complete the task – the task of copying faithfully the whole Book of Nature, and of writing the new science.

Unfortunately, Bacon did not get the research grant for which he was looking. The great Foundations did not yet exist and as a consequence, sad to say, the science of nature is still unfinished.

Bacon's naïve and amateurish optimism was a source of encouragement and inspiration for the great scientific amateurs who founded the Royal Society, modelling it after the central research institution envisaged by Bacon in his *New Atlantis*.

Bacon was the prophet, the great inspirer of the new religion of science, but he was not a scientist. Yet the inspiration and the influence of his new theology of nature were at least as great and as lasting as those of his contemporary Galileo, who might be described as the true founder of modern experimental science. More especially, Bacon's naïve view concerning the essence of natural science, and the distinction or demarcation drawn by him between the new natural science on the one hand and the old theology and philosophy on the other, became the main dogma of the new religion of science. It is a dogma to which scientists as well as philosophers have tenaciously adhered down to our own day. And it is only in recent years that some scientists have become willing to listen to those who criticize this dogma.

The Baconian dogma I have in mind asserts the supreme merits of observation and the viciousness of theorizing speculation. I shall call this dogma, briefly, by the name 'observationism'.

According to Bacon, the nature or essence of the method of the new science of nature, the method which distinguishes and demarcates it from the old theology and from metaphysical philosophy, can be explained as follows:

Man is impatient. He likes quick results. So he jumps to conclusions.

This is the old, the vicious, the speculative method. Bacon called it 'the method of anticipations of the mind'. It is a false method, for it leads to prejudices. (The term 'prejudice' was coined by Bacon.)

Bacon's new method, which he recommends as the true way to knowledge, and also as the way to power, is this. We must purge our minds of all prejudices, of all preconceived ideas, of all theories – of all those superstitions, or 'idols', which religion, philosophy, education, or tradition may have imparted to us. When we have thus purged our minds of prejudices and impurities, we may approach nature. And nature will not mislead us. For it is not nature that misleads us but only our own prejudices, the impurities of our own minds. If our minds are pure, we shall be able to read the Book of Nature without distorting it: we have only to open our eyes, to observe things patiently, and to write down our observations carefully, without misrepresenting or distorting them, and the nature or essence of the thing observed will be revealed to us.

This is Bacon's method of observation and induction. To put it in a nutshell: pure untainted observation is good, and pure observation cannot err; speculation and theories are bad, and they are the source of all error. More especially, they make us misread the Book of Nature – that is, misinterpret our observations.

Bacon's observationism and his hostility to all forms of theoretical thought were revolutionary, and were felt to be so. They became the battle cry of the new secularized religion of science, and its most cherished dogma. This dogma had an almost unbelievable influence upon both the practice and the theory of science, and this influence is still strong in our own day.

In order to show that this dogma did not express the general belief of scientists contemporary with Bacon, I shall once again briefly contrast Bacon with Galileo.

Bacon, the philosopher of science, was, quite consistently, an enemy of the Copernican hypothesis. Don't theorize, he said, but open your eyes and observe without prejudice, and you cannot doubt that the Sun moves and that the Earth is at rest.

Galileo, the great scientist and defender of the Copernican 'System of the World', paid homage to Aristarchus and Copernicus precisely because they were bold enough to produce speculative theories which not only go beyond, but also contradict, all that we believe ourselves to know from observation.

I may perhaps quote a passage from Galileo's *Dialogue Concerning the Two Chief World Systems*:[1]

> I shall never be able to express strongly enough my admiration for the greatness of mind of these men who conceived this [heliocentric] hypothesis and held it to be true. In violent opposition to the evidence of their own senses and by sheer force of intellect, they preferred what reason told them to that which sense experience plainly showed them ... I repeat, there is no limit to my astonishment when I reflect how Aristarchus and Copernicus were able to let reason conquer sense, and in defiance of sense make reason the mistress of their belief.

This is Galileo's testimony to the way in which bold and purely speculative scientific theories may free us from our prejudices. Bacon, on the contrary, held that these new theories were speculative prejudices, that theoretical thinking always creates prejudices, that only its abandonment can help us to free ourselves from prejudices, and that thought can never achieve this.

Before turning to criticize the Baconian dogma, and to replace it by a very different view of experimental as well as theoretical science, I wish to add a final remark about Bacon.

Bacon, I suggest, was not a scientist but a prophet. He was a prophet not only in the sense that he propagated the idea of an experimental science, but also in the sense that he foresaw, and inspired, the industrial revolution. He had the vision of a new age, of an industrial age which would also be an age of science and of technology. Referring to the accidental discovery of gunpowder, and of silk, he spoke of the possibility of a systematic scientific search for other useful substances and materials, and of a new society in which, through science, men would find salvation from misery and poverty. Thus the new religion of science held a new promise of heaven on earth – of a better world which, with the help of new knowledge, men would create for themselves. Knowledge is power, Bacon said, and this idea, this dangerous idea, of man's mastery over nature – of men like gods – has been one of the most influential of the ideas through which the religion of science has transformed our world.

II

I shall now very briefly criticize Bacon's anti-theoretical dogma and his view of science, and then turn to my own view of science – and in particular of experimental science – which I propose to put in its place.

1   The idea that we can purge our minds of prejudices at will and so get rid of all preconceived ideas or theories, prior to, and preparatory to, scientific discovery, is naïve and mistaken. It is mainly through scientific discovery that we learn that certain of our ideas – such as those of the flat earth or the moving sun – are prejudices. We discover the fact that one of the beliefs we held was a prejudice only after the advance of science has led us to discard it. For there is no criterion by which we could recognize prejudices in anticipation of this advance.

2   The rule 'Purge yourself of prejudice!' can therefore have only the dangerous result that, after having made an attempt or two, you may think that you have succeeded – with the result, of course, that you will stick more tenaciously to your prejudices and dogmas, especially to those of which you are unconscious.

3   Moreover, Bacon's rule was 'purge your mind of *all* theories!' But a mind so purged would not only be a pure mind: it would be an empty mind.

4   We always operate with theories, even though more often than not we are unaware of them. The importance of this fact should never be played down. Rather, we should try, in each case, to formulate explicitly the theories we hold. For this makes it possible to look out for alternative theories, and to discriminate critically between one theory and another.

5   There is no such thing as a 'pure' observation, that is to say, an observation without a theoretical component. All observation – and especially all experimental observation – is an interpretation of facts in the light of some theory or other.

This last remark leads me to a crucial point – the point which I should be inclined to call 'Bacon's problem'. It is this.

6   Bacon was aware of the general tendency to interpret observed facts in the light of theories, and he was keenly awake to the very

real dangers of this tendency. He saw that if we interpret the observed facts in the light of preconceived theories or 'prejudices', then we are liable to confirm, and to strengthen these prejudices by our observations, whatever the actual facts may be. Thus prejudices make it impossible for us to learn from experience: they form an impassable barrier to the advancement of science through observation and experiment.

The point is so important that it should be illustrated by some examples.

What Bacon had in mind was something like this. Let a man hold some religious creed – say, the Zoroastrian or Manichaean heresy which sees our world as an arena of conflict between a good and an evil power. Then all his observations will only confirm his belief. In other words, he will never be able to correct it by experience, or to learn from experience.

There is a modern secular parallel to this theological example. Take a man who believes in the theory that all history is a history of class struggle, and that modern history is the history of the struggle between virtuous proletarians and vile capitalists. If he holds this belief, then whatever he observes or experiences and whatever the newspapers report or fail to report will be interpreted by him in terms of this belief, and will therefore tend to reinforce it.

Or take a third example. Psychoanalysts tend to speak of what they call their 'clinical observations', and of the fact that these observations invariably support the psychoanalytic theory. These clinical observations are, however, always interpreted: they are interpreted in accordance with established psychoanalytic theory. This raises the question: Is it legitimate to claim that the observations support the theory? Or to put it in another way: Can we conceive of any human behaviour that we could not interpret in psychoanalytic terms? If the answer to this question is 'no', then we can say, prior to any observation, that every conceivable observation will be interpretable in the light of psychoanalytic theory and that it will, thereby, appear to support it. But if this can be said prior to any observation, then this kind of support must not be described as genuinely empirical or observational.

This, I suppose, is the difficulty that Bacon felt. The only escape from it that he could devise was the impracticable proposal to purge our minds of all theories, and adhere to 'pure' observation.

III

With this, I will now leave Bacon's views in order to give you my own view of the matter. I will first propose a simple solution of Bacon's problem.

My solution consists of two steps.

First, every scientist who claims that his theory is supported by experiment or observation should be prepared to ask himself the following question: Can I describe any possible results of observation or experiment which, if actually reached, would refute my theory?

If not, then my theory is clearly not an empirical theory. For if all conceivable observations agree with my theory, then I cannot be entitled to claim of any particular observation that it gives empirical support to my theory.

Or in short, only if I can say how my theory might be refuted, or falsified, can I claim that my theory has the character of an empirical theory.

This criterion of demarcation between empirical and non-empirical theories I have also called the criterion of falsifiability or the criterion of refutability. It does not imply that irrefutable theories are false. Nor does it imply that they are meaningless. But it does imply that, as long as we cannot describe what a possible refutation of a certain theory would be like, that theory may be regarded as lying outside the field of empirical science.

The criterion of refutability or falsifiability may also be called the criterion of testability. For testing a theory, like testing a piece of machinery, means trying to fault it. Thus a theory which we know in advance cannot possibly be faulted or refuted is not testable.[2]

It should be made quite clear that there are many examples in the history of science of theories which at some stage of the development of science were not testable but which became testable at a later stage. An obvious example is atomic theory. An example within modern physical theory which would deserve a detailed discussion is the theory of the neutrino.

When this theory was first proposed by Pauli, it was clearly not testable. It was even said, at one time, that the neutrino is so defined that the theory cannot be tested. About thirty years later the theory was not only found to be testable, but to pass its test with flying colours. This should be a warning to those who are inclined to say that nontestable theories are meaningless (a view which has often but mistakenly been attributed to me) or that they have no 'cognitive significance'.

So much for the criterion of the empirical character of a theory. It does not completely solve Bacon's problem. But it allows us to reject many of those unjustifiable claims to observational support which so worried Bacon.

The criterion of refutability, or falsifiability, or testability, is only the first step in the solution of Bacon's problem. As we have seen, this step is taken by asking a scientist who claims that his theory is supported by experiment or observation, 'Is your theory refutable? And what experiment or observation would you accept as a refutation?'

If the answers to these questions are satisfactory, then, and only then, can we proceed to take the second step in our solution to Bacon's problem. It amounts to this.

Observations or experiments can be accepted as supporting a theory (or a hypothesis, or a scientific assertion) only if these observations or experiments are severe tests of the theory – or, in other words, only if they result from serious attempts to refute the theory, and especially from trying to find faults where these might be expected in the light of all our knowledge, including our knowledge of competing theories.

I believe that this, in principle, solves Bacon's problem.

The solution amounts to this. Agreement between theory and observation should count for nothing unless the theory is testable, and unless the agreement is found as the result of serious attempts to test it. But testing a theory means trying to find its weak spots. It means trying to refute it. And a theory is testable only if it is (in principle) *refutable*.

IV

Let us look at a few examples. Psychoanalysis would become refutable only if it denied that certain possible or conceivable forms of human behaviour do, in fact, occur.

Newton's theory of gravity is highly testable, for example, because its theory of perturbances predicts certain deviations from Kepler's planetary orbits, and this prediction may be refuted. Einstein's theory of gravity is highly testable because it predicts certain deviations from Newton's planetary orbits, and this prediction may be refuted. It also predicts the curvature of light rays and the retardation of atomic clocks in strong gravitational fields, and again these predictions may be refuted.

There is a difficulty with Darwinism. While Lamarckism appears to be not only refutable but actually refuted (because the kind of acquired adaptations which Lamarck envisaged do not appear to be hereditary), it is far from clear what we should consider a possible refutation of the theory of natural selection. If, more especially, we accept that statistical definition of fitness which defines fitness by actual survival, then the theory of the survival of the fittest becomes tautological, and irrefutable.

Darwin's great achievement was this, I believe. He showed that what appeared to be purposeful adaptation may be explained by some mechanism – such as, for example, the mechanism of natural selection. This was a tremendous achievement. But once it is shown that a mechanism of this kind is possible, we ought to try to construct alternative mechanisms, and then try to find some crucial experiments to decide between them, rather than foster the belief that the Darwinian mechanism is the only possible one.

Or let us take as an example a theory more closely related to experimental work: the theory of synaptic transmission. The chemical theory of transmission (as against the competing electrical theory) passed a severe test when acetylcholine was artificially applied to the contact region of the muscle fibre. The fact that it triggered the impulse like a firing nerve could be claimed in support of the chemical theory.[3]

The view here presented may be summed up by saying that the decisive function of observation and experiment in science is criticism. Observation and experiment cannot establish anything conclusively, for there is always the possibility of a systematic error through systematic misinterpretation of some fact or other. But observation and experiment certainly play an important part in the critical discussion of scientific theories. Essentially, they help us to eliminate the weaker theories. In this way they lend support, though only for the time being, to the surviving theory – that is, to the theory which has been severely tested but not refuted.

V

The modern view of science – the view that scientific theories are essentially hypothetical or conjectural, and that we can never be sure that even the best established theory may not be overthrown and replaced by a better approximation – is, I believe, the result of the Einsteinian revolution.

For there never was a more successful theory, or a better tested theory, than Newton's theory of gravity. It succeeded in explaining both terrestrial and celestial mechanics. It was most severely tested in both fields for centuries. The great physicist and mathematician Henri Poincaré not only believed that it was true – this of course was everybody's belief – but that it was true by definition, and that it would therefore remain the invariable basis of physics to the end of man's search for truth. And Poincaré believed this in spite of the fact that he actually anticipated – or that he came very close to anticipating – Einstein's special theory of relativity. I mention this in order to illustrate the tremendous authority of Newton's theory down to the very last.

Now the question whether or not Einstein's theory of gravity is an improvement upon Newton's, as most physicists think it is, may be left open. But the mere fact that there was now an alternative theory which explained everything that Newton could explain and, in addition, many more things, and which passed at least one of the crucial tests which Newton's theory seemed to fail, destroyed the unique place held by Newton's theory in its field. Newton's theory was thus reduced to the status of an excellent and successful conjecture, a hypothesis competing with others, and one whose acceptability was an open question. Einstein's theory thus destroyed the authority of Newton's, and with it something of even greater importance – authoritarianism in science.

Those of you who are my contemporaries may remember the days when complete authority was claimed by the secular religion of science. Hypotheses were recognized as playing a role in science, but their role was heuristic and transitory: science itself was believed to be a body of knowledge. It did not consist of hypotheses, but of proved theories – proved theories like that of Newton.

It is interesting in this context that Max Planck tells the story that, when he was an ambitious young man, a famous physicist tried to discourage him from studying physics with the remark that physics was about to reach its ultimate completion, and that there were no longer any great discoveries to be made in this field.

This period of authoritarian science has passed, I suppose forever, owing to the Einsteinian revolution.[4] It is interesting in this connection to note that Einstein himself did not hold that his general theory was true – though he did believe that it was a better approximation to the truth than Newton's, and that a still better approximation, and of course also the true theory (if ever found), would have to contain, in their turn, general relativity as an approximation. In other words, Einstein was clear from the very first about the essentially conjectural character of his theories.

As I have said earlier, it was part of the religion of science before Einstein to claim authority for science. Admittedly, there were a few heretics, notably the great American philosopher Charles S. Peirce, who said before Einstein that science shared the fallibility of all human endeavours. Yet Peirce's fallibilism became influential mainly after the Einsteinian revolution.

VI

I have mentioned these historical facts merely because I wish to stress that the change from the authoritarian theory of scientific knowledge to an anti-authoritarian and critical theory is quite recent. This also explains why the view that the method of science is essentially the method of critical discussion, and of a critical examination of competing conjectures or hypotheses, is still felt by many to be inappropriate to the experimental sciences – why so many people still feel that what is based upon careful laboratory work has more than merely hypothetical status.

To combat this view, I may choose an example from chemistry. If you had asked an experimental chemist, before the discovery of heavy water, what branch of chemistry was most secure – least likely to be overthrown or corrected by new revolutionary discoveries – he would almost certainly have said, the chemistry of water. In fact, water was used in the definition of one of the fundamental units of physics, the gram, forming part of the centimetre–gram–second system. And hydrogen and oxygen were used as the theoretical and practical basis in the determination of all atomic weights.

All this was completely upset by the unexpected discovery of heavy water, from which we may learn the lesson that we can never know which part of science will have to be revised next.

Or take a still more recent example from physics: the breakdown of parity. This was one of those cases in which it turned out after the event that there had been many observations – photographs of particle tracks – from which we might have read off the result, but that the observations had been either ignored or misinterpreted. Much the same thing had happened before when the positron was discovered, and before this when the neutron was discovered. Earlier still, before the discovery of X-rays, it happened to William Crookes himself, the inventor of the Crookes tube with the help of which X-rays were subsequently discovered.

VII

I may now perhaps sum up the first part of my talk by restating all the controversial things I have been saying in a number of theses which I shall try to put in as challenging a form as I can.

1 All scientific knowledge is hypothetical or conjectural.
2 The growth of knowledge, and especially of scientific knowledge, consists in learning from our mistakes.
3 What may be called the method of science consists in learning from our mistakes systematically: first, by taking risks, by daring to make mistakes – that is, by boldly proposing new theories; and secondly, by searching systematically for the mistakes we have made – that is, by the critical discussion and the critical examination of our theories.
4 Among the most important arguments which are used in this critical discussion are arguments from experimental tests.
5 Experiments are constantly guided by theory, by theoretical hunches of which the experimenter is often not conscious, by hypotheses concerning possible sources of experimental errors, and by hopes or conjectures about what will be a fruitful experiment. (By *theoretical* hunches I mean guesses that experiments of a certain kind will be theoretically fruitful.)
6 What is called scientific objectivity consists solely in the critical approach: in the fact that if you are biased in favour of your pet theory, some of your friends and colleagues (or failing these, some workers of the next generation) will be eager to criticize your work – that is to say, to refute your pet theories if they can.
7 This fact should encourage you to try to refute your own theories yourself – that is to say, it may impose some discipline upon you.
8 In spite of this, it would be a mistake to think that scientists are more 'objective' than other people. It is not the objectivity or detachment of the individual scientist but of science itself (what may be called 'the friendly-hostile cooperation of scientists' –

that is, their readiness for mutual criticism) which makes for objectivity.

9   There is even something like a methodological justification for individual scientists to be dogmatic and biased. Since the method of science is that of critical discussion, it is of great importance that the theories criticized should be tenaciously defended. For only in this way can we learn their real power. And only if criticism meets resistance can we learn the full force of a critical argument.

10 The fundamental role played in science by theories or hypotheses or conjectures makes it important to distinguish between testable (or falsifiable) and non-testable (or non-falsifiable) theories.

11 Only a theory which asserts or implies that certain conceivable events will not, in fact, happen is testable. The test consists in trying to bring about, with all the means we can muster, precisely these events which the theory tells us cannot occur.

12 Thus, every testable theory may be said to forbid the occurrence of certain events. A theory speaks about empirical reality only in so far as it sets limits to it.

13 Every testable theory can thus be put into the form 'such and such cannot happen'. For example, the second law of thermodynamics can be formulated as saying that a perpetual motion machine of the second kind cannot exist.

14 No theory can tell us anything about the empirical world unless it is in principle capable of clashing with the empirical world. And this means, precisely, that it must be refutable.

15 Testability has degrees: a theory which asserts more, and thus takes greater risks, is better testable than a theory which asserts very little.

16 Similarly, tests can be graded as being more or less severe. Qualitative tests, for example, are in general less severe than quantitative tests. And tests of more precise quantitative predictions are more severe than tests of less precise predictions.

17 Authoritarianism in science was linked with the idea of establishing, that is to say, of proving or verifying, its theories. The critical approach is linked with the idea of testing, that is to say, of trying to refute, or to falsify, its conjectures.

VIII

I now turn to the second part of this talk, devoted to *problems* and their role in science.

Science begins with observation, says Bacon, and this saying is an integral part of the Baconian religion. It is still widely accepted, and still repeated *ad nauseam* in the introduction to even some of the best textbooks in the field of the physical and biological sciences.

I propose to replace this Baconian formula by another one.

Science, we may tentatively say, begins with theories, with prejudices, superstitions, and myths. Or rather, it begins when a myth is challenged and breaks down – that is, when some of our expectations are disappointed. But this means that *science begins with problems*, practical problems or theoretical problems.

Before going on to develop my thesis here more fully, I may perhaps say a few words about the term 'expectation', which I have just used.

In walking down steps it sometimes happens that we suddenly discover that we expected another step (which was not there) or, on the contrary, that we expected no other step (while in reality there was one). The unpleasant discovery that we were mistaken makes us realize that we had certain unconscious expectations. And it shows us that there are thousands of such unconscious expectations. A similar example is this: if we sit and work in a room in which a clock can be heard ticking, we may hear that the clock has suddenly stopped. This makes us conscious of the fact that we expected it to go on ticking – even though we may not have been conscious of hearing it.

A study of animal behaviour teaches us that animals similarly adjust their behaviour to impending events and become disturbed if the expected event does not happen.

We may say that an expectation, conscious or unconscious, corresponds, on the pre-scientific level, to what we call, on the scientific level, 'a conjecture' (about an impending event), or 'a theory'.

In my views about the methods of science and especially about the role of observation[5] I disagree with almost everybody except Charles Darwin and Albert Einstein. Einstein, incidentally, explained his views on these matters concisely in his Herbert Spencer Lecture, delivered in Oxford in 1933, and entitled *On the Methods of Theoretical Physics*.[6] There he told his audience not to believe those scientists who say that their methods are inductive.

Since, as I said, I disagree about these matters with almost everybody, I cannot hope that I shall convince you and I shall not try to do so. All I shall attempt is to draw your attention to the fact that there are some people who hold views on these matters differing widely from the usual ones, and that men like Darwin and Einstein are among them.

My thesis, as I have already indicated, is that we do not start from observations but always from problems: from practical problems, or from a theory which has run into difficulties – that is to say, a theory which has raised, and disappointed, certain *expectations*.

Once we are faced with a problem, we proceed by two kinds of attempt. We attempt to guess, or to conjecture, a solution to our problem. And we attempt to criticize our usually somewhat feeble solutions. Sometimes a guess or a conjecture may withstand our criticism and our experimental tests for quite some time. But as a rule, we find that our conjectures can be refuted, or that they do not solve our problem, or that they solve it only in part. And we find that even the best solutions – those able to resist the most severe criticism of the most brilliant and ingenious minds – soon give rise to new difficulties, to new problems. Thus we may say that our knowledge grows as we proceed from old problems to new problems by means of *conjectures and refutations* – by the refutation of our theories or, more generally, of our *expectations*.

I suppose that some of you will agree that we usually start from problems. But you may still think that our problems must have been the result of observation and experiment, since prior to receiving impressions through our senses, our mind is a *tabula rasa*, an empty slate, a blank – for there can be nothing in our intellect which has not entered it through our senses.

But it is just this venerable idea that I am combating. I assert that every animal is born with many, usually unconscious, expectations – in other words with something closely corresponding to hypotheses, and thus to hypothetical knowledge. And I assert that we always have, in this sense, inborn knowledge to start from, even though it may be quite unreliable. This inborn knowledge, these inborn expectations, will, if disappointed, create our first problems. And the ensuing growth of knowledge may therefore be described as consisting throughout of corrections and modifications of previous knowledge – of previous expectations or hypotheses.

Thus I am turning the tables on those who think that observation must precede expectations and problems. And I even assert that observation cannot, for logical reasons, be prior to all problems, although, obviously, it will sometimes be prior to some problems – for example to those that spring from an observation which has disappointed some of our expectations or which has refuted some of our theories.

Now this fact – that observation cannot precede all problems – may be illustrated by a simple experiment which I wish to carry out, by your leave, with yourselves as experimental subjects. My experiment is to ask you to observe, here and now. I hope you are all cooperating and observing! Yet I fear that some of you, instead of observing, will feel a strong urge to ask: 'What do you want me to observe?'

If this is your response, then my experiment was successful. For what I am trying to illustrate is that, in order to observe, we must have in mind a definite question which we might be able to decide by observation. Charles Darwin knew this when he wrote: 'How odd it is that anyone should not see that all observation must be for or against some view …'.[7]

I cannot, as I said before, hope to convince you of the truth of my thesis that observation comes after expectation or hypothesis. But I do hope that I have been able to show you that there may exist an alternative to the venerable doctrine that knowledge – especially scientific knowledge – starts from observation. (The still more venerable doctrine that all knowledge starts from perception or sensation or sense-data, which, of course, I also reject, is, incidentally, at the root of the fact that 'problems of perception' are still widely considered to form a respectable part of philosophy or, more precisely, of epistemology.)

IX

Now let us look a little more closely at the way in which we get acquainted with a problem.

We start, I say, with a problem – a difficulty. It is perhaps a practical problem, or a theoretical problem. Whatever it may be, when we first encounter the problem we cannot, obviously, know much about it. At best, we have only a vague idea what our problem really consists of. How, then, can we produce an adequate solution? Obviously, we cannot. We must first get better acquainted with the problem. But how?

My answer is very simple: by producing a very inadequate solution, and by criticizing this inadequate solution. Only in this way can we come to understand the problem. For to understand a problem means to understand why it is not easily soluble – why the more obvious solutions do not work. We must therefore produce these obvious solutions and try to find out why they will not do. In this way, we become *acquainted* with the problem. And in this way we may proceed from bad solutions to slightly better ones – provided always that we have the ability to guess again.

A very trivial example of this method of *attempting to solve a problem by trial and the elimination of error* is the task of dividing a largish number – say 22376 – by another one – say 2784. Our usual method is to *guess* the first figure of the quotient – our guess may be that it is 7 – and to try out whether our guess was correct. If our guess was 7, we easily find that we were in error, and that we have to replace 7 by 8. There are many less trivial mathematical problems for which the standard method of solving them is to start with a guess and subsequently to correct the error made.[8]

These examples should make it clear that the method of trial and error-elimination is utterly different from the so-called (but in my view nonexistent) 'method of induction by repetition'. Nevertheless, the two have often been confused.

In simple mathematical problems the solution can always be found after a small number of trials and errors, or even after only one. But this is not, of course, generally true of mathematical problems (some of which are insoluble). And it is certainly not true of problems in the empirical sciences. Yet it is generally true that the best if not the only method of learning something about a problem is to try first to solve it by guessing and then to try to pinpoint the mistakes we have made.[9]

This, I think, is what is meant by 'working on a problem'. And if we have worked on a problem long enough, and intensively enough, we begin to know it, to understand it, in the sense that we know what kind of solution does not do at all (because it simply misses the point of the problem) and what kind of requirement would have to be met by a serious attempt at a solution. In other words, we begin to see the ramifications of the problem, its sub-problems, and its connection with other problems.

At this stage our tentative solutions may be submitted to the criticism of others – to critical discussion, that is – and perhaps even be published.

Or if you are an experimentalist, you may now proceed to test your solution. If it is the solution of a practical problem of experimentation, you will try it out in various experiments. If it is a conjecture, a hypothesis, you will test it with the help of experiments.

These experimental tests are, of course, again part of the process of critically 'working on a problem': of getting to know it, of getting acquainted and really familiar with it, and thus perhaps of improving one's chances of finding, some day, a satisfactory and illuminating solution.

However this may be, the really important point I wish to make is this. If the question is asked, 'What is it to understand a problem?', my answer is that there is only one way to learn to understand a serious problem – whether it is now purely theoretical or a practical problem of experimentation. And this is to try to solve it, and to fail. Only if we find that some facile and obvious solution does not solve our problem do we begin to understand it. For a problem is a difficulty. Understanding it means experiencing this difficulty. And this can only be done by finding out that there is no easy and obvious solution to it.

Thus we become acquainted with a problem only when we have many times tried in vain to solve it. And after a long series of failures – of attempts yielding solutions which turn out to be unacceptable – we may even become experts in this particular problem. We shall have become experts in the sense that, whenever somebody else offers a new solution – for example, a new theory – it will be either one of those solutions which we have tried out in vain (so that we shall be able to explain why it does not work) or it will be a new solution. In this case we may be able to find out quickly whether or not it gets over at least those standard difficulties which we know so well from our unsuccessful endeavours.

My point is that even if we persistently fail to solve our problem, we shall have learned a great deal by having wrestled with it. The more we try, the more we shall learn about it – even if we fail every time. It is clear that, having become in this way utterly familiar with a problem – that is, with its difficulties – we may have a better chance to solve it than somebody who does not even understand the difficulties. But it is all a matter of chance: in order to solve a difficult problem one needs not only understanding but also luck.

Thus, like science itself, which begins and ends with problems and progresses through wrestling with them, the individual scientist should also begin and end with his problem and wrestle with it. Moreover, while wrestling with it, he will not merely learn to understand the problem, but he will actually change it. A change of emphasis may make all the difference – not only to our under-standing, but to the problem itself, to its fertility and significance, and to the prospects of an interesting solution. It is important for a scientist to be awake to these changes and shifts, and not to make them either unconsciously or surreptitiously. For it often happens that a reformulation of a problem can reveal to us almost the whole of its solution.

X

My view of the significance of problems for the methodology or theory of scientific knowledge may perhaps be summed up by the following considerations.

The theory of knowledge – and especially the theory of scientific knowledge – is constantly faced with a near paradox which may be brought home to us by the clash of the following two theses.

First thesis: Our knowledge is vast and impressive. We know not only innumerable details and facts of practical significance, but also many theories and explanations which give us an astonishing intellectual insight into dead and living subjects, including ourselves, and human societies.

Second thesis: Our ignorance is boundless and overwhelming. Every new bit of knowledge we acquire serves to open our eyes further to the vastness of our ignorance.

Both of these theses are true, and their clash characterizes our knowledge-situation. The tension between our knowledge and our ignorance is decisive for the growth of knowledge. It inspires the advance of knowledge, and it determines its ever-moving frontiers.

The word 'problem' is only another name for this tension – or rather, a name denoting various concrete instances of it.

As I suggested above, a problem arises, grows, and becomes significant through our failures to solve it. Or to put it another way, the only way of getting to know a problem is to learn from our mistakes.

This applies to pre-scientific knowledge and to scientific knowl-edge.

My view of the method of science is, very simply, that it systematizes the pre-scientific method of learning from our mistakes. It does so by the device called critical discussion.

My whole view of scientific method may be summed up by saying that it consists of these three steps:

1   We stumble over some problem.
2   We try to solve it, for example by proposing some theory.
3   We learn from our mistakes, especially from those brought home to us by the critical discussion of our tentative solutions – a discussion which tends to lead to *new problems*.

Or in three words: *problems – theories – criticism*.

I believe that in these three words the whole procedure of rational science may be summed up.[10]

XI

Having discussed problems and their growth at some length, I now turn to theories. I shall discuss the question: What is meant by saying that we '*understand*' a scientific theory?

This question has been much discussed, and it has been suggested that we should not speak at all about 'understanding' theories – that the idea that we can understand a theory is out of date. It has also been suggested that those who, like myself, speak about 'under-standing' mean either the understanding of a crude mechanism, like a clock, or else 'understanding' in the sense of being able to draw a picture, or make a model, of the process in question. And it is then pointed out, and quite correctly, that modern physical theory no longer confines itself to clockwork mechanisms, or to picturable processes. From this it is concluded – wrongly, I believe – that the whole idea of 'understanding' a theory is out of date. And this conclusion is widely accepted, not only by physicists but also by some biologists.

I do not think that the conclusion is correct. And I do not see any reason why understanding a theory should be any more out of date than understanding a problem: a process which I have described without appealing to models or pictures which may be intuited or visualized.

Understanding a theory, I suggest, means understanding it *as an attempt to solve a certain problem*. This is an important proposition, and one which too few people understand.

What is the point of, say, Newton's theory? It is an attempt to solve the problem of explaining Kepler's and Galileo's laws. Without understanding the problem situation which gave rise to the theory, the theory is pointless – that is, it cannot be understood.

Or take as an example Bohr's theory (1913) of the hydrogen atom. This theory was describing a model, and was therefore intuitive and visualizable. Yet it was also very perplexing. Not because of any intuitive difficulty, but because it assumed, contrary to Maxwell's and Lorentz's theory and to well-known experimental effects, that a periodically moving electron, a moving electric charge, need not always create a disturbance of the electromagnetic field, and so need not always send out electromagnetic waves. This difficulty is a logical one – a clash with other theories. And no one can be said to understand Bohr's theory who does not understand this difficulty and the reasons why Bohr boldly accepted it, thus departing in a revolutionary way from earlier and well-established theories.

But the only way to understand Bohr's reasons is to understand his problem – the problem of combining Rutherford's atom model with a theory of emission and absorption of light, and thus with Einstein's photon theory, and with the discreteness of atomic spectra. The understanding of Bohr's theory does not lie in visualizing it intuitively but in gaining familiarity with the problems it tries to solve, and in the appreciation of both the explanatory power of the solution and the fact that the new difficulty which it creates constitutes an entirely new problem of great fertility.

The question whether or not a theory or a conjecture is more or less satisfactory or, if you like, *prima facie* acceptable as a solution of the problem which it sets out to solve is largely a question of purely deductive logic. It is a matter of getting acquainted with the logical conclusions which may be drawn from the theory, and of judging whether or not these conclusions (a) yield the desired solution and (b) yield undesirable by-products – for example some insoluble paradox, some absurdity.

XII

It may be appropriate at this stage to say something about the acceptance of theories – much discussed by philosophers of science as the question of 'verification'.

To begin with, I wish to make it quite clear that I regard the question of the acceptance of a theory or conjecture as one whose importance is much over-rated (quite apart from the fact that I don't believe in the verification or verifiability of theories, but this I shall not discuss here).[11]

Consider just one example. Einstein proposed his theory of general relativity, he defended it patiently against violent criticism, he suggested that it was an important advance, and that it should be accepted as an improvement on Newton's theory – but he never accepted it himself in that sense of 'accepted' which almost all philosophers of science regard as important. What I mean is this. Philosophers of science speak as if there were a body of knowledge, called science, which consists, in the main, of accepted theories. But this seems to me utterly mistaken, and a residue of the dreams of authoritarian science prevailing in the days when people thought that we were just on the verge of completing the task of science, a thing that Bacon believed in 1600, and that some competent physicists still believed in 1900, as Max Planck has told us.

It seems to me that most philosophers of science use the term 'accepted' or 'acceptable' as a substitute for 'believed in' or 'worthy of being believed in'. There may be a lot of theories in science which are true and therefore worthy of being believed in. But according to my view of the matter, this worthiness is no concern of science. For science does not attempt positively to justify or to establish this worthiness. On the contrary, it is mainly concerned with criticizing it. It regards, or should regard, the overthrow of even its most admirable and beautiful theories as a triumph, an advance. For we cannot overthrow a good theory without learning an immense amount from it and from its failure. As always, we learn from our mistakes.

The overthrow of a theory always creates new problems. But even if a new theory is not yet overthrown, it will, as we have seen from the example of Bohr's theory, create new problems. And the quality, the fertility, and the depth of the new problems which a theory creates are the best measures of its intrinsic scientific interest.

To sum up, the question of the acceptance of theories should, I propose, be demoted to the status of a minor problem. For science may be regarded as a growing system of problems, rather than as a system of beliefs. And for a system of problems, the tentative acceptance of a theory or a conjecture means hardly more than that it is considered worthy of further criticism.

XIII

I have not said anything about induction so far, and I should not have said anything were I not afraid to disappoint some of those who came to hear a philosopher on scientific method – and thus on induction.

So I must say now that I do not believe there is such a thing as an inductive method or an inductive procedure – unless indeed you decide to use the name 'induction' for that method of critical discussion and of attempted refutations which I have described here.

I never quarrel about words, and I have of course no serious objection if you wish to call the method of critical discussion 'induction'. But if you do, then you should be aware of the fact that it is very different from anything that has ever been called 'induction' in the past. For induction was always supposed to *establish* a theory, or a generalization, while the method of critical discussion does not establish anything. Its verdict is always and invariably 'not proven'. The best that it can do – and this it does rarely – is to come out with the verdict that a certain theory appears to be the best available (that is to say, the best so far submitted to examination and discussion), that it appears to solve much of the problem it was designed to solve, and that it has survived the severest tests that we were able to devise. But this does not, of course, establish the theory as true (that is to say, as corresponding to the facts, or as an adequate description of reality) – although we may say that what such a positive verdict amounts to is that, in the light of our critical discussion, the theory appears to be the best approximation to the truth so far attained.[12]

In fact, the idea of 'better approximation to the truth' is at once the main standard of our critical discussion and an aim we hope to attain as a result of that discussion. Among our other standards is the explanatory power of a theory, and its simplicity.[13]

In the past, the term 'induction' has been used mainly in two senses. The first is repetitive induction (or induction by enumeration). This consists of often repeated observations and experiments, which are supposed to serve as premises in an argument establishing some generalization or theory. The invalidity of this kind of argument is obvious: no amount of observation of white swans establishes that all swans are white (or that the probability of finding a non-white swan is small). In the same way, no amount of observed spectra of hydrogen atoms on earth establishes that all hydrogen atoms emit spectra of the same kind. Theoretical considerations, however, may suggest the latter generalization, and further theoretical considerations may suggest that we should modify it by introducing Doppler shifts and Einsteinian gravitational red-shifts.

Thus *repetitive induction is out*: it cannot establish anything.

The second main sense in which the term 'induction' has been used in the past is eliminative induction – induction by the method of eliminating or refuting false theories. This may look at first sight very much like the method of critical discussion that I am advocating. But in fact it is very different. For Bacon and Mill and other exponents of this method of eliminative induction believed that by eliminating all false theories we can finally *establish* the true theory. In other words, they were unaware of the fact that the number of competing theories is always infinite – even though there are as a rule at any particular moment only a finite number of theories before us for consideration. I say 'as a rule', for sometimes an infinite number is before us. For example, it was suggested that we should modify Newton's inverse square law of attraction, replacing the square by a power differing slightly from the number 2. This proposal amounted to the suggestion that we should consider an infinite number of slightly different corrections to Newton's law.

The fact that there is always an infinity of logically possible solutions to every problem is a decisive fact for the philosophy of science. It is one of those things that makes science such a thrilling adventure. For it renders inefficient all merely routine methods. It means that scientists must use imagination and bold ideas, though always tempered by severe criticism and severe tests.

It also shows, incidentally, the mistake of those who think that the aim of science is merely to establish correlations between observed events, or observations (or, worse, 'sense data'). What we aim at in science is much more. We aim at discovering new worlds behind the world of ordinary experience: such as, perhaps, a microscopic or submicroscopic world – gravitational, chemical, electrical, and nuclear forces, some of them, perhaps, reducible to others, and others not. It is the discovery of these new worlds, of these new undreamt-of possibilities, which adds so much to the liberating power of science. Correlation coefficients are not interesting if they merely correlate our observations. They are interesting only if they help us to learn more about these worlds.

105

XIV

Let me conclude this part of my talk with a practical proposal.

There is a tradition still alive in the writing of scientific papers which I have dubbed '*the inductive style*'. I am sure you all know it, and some of you may still practise it. A very well-known form of it is to write a paper by first describing the experimental arrangements, then the observations, possibly a curve which may link them up, and perhaps concluding (in small print) with a hypothesis. This inductive or Baconian style has a long and glorious history: great and world-shaking papers have been written in this style – for example, Sir Alexander Fleming's paper reporting his first observations of penicillin.

But we all know that Fleming did not merely observe effects: he knew many things beforehand. He knew about Ehrlich's hopes, and the possibility of antibiotic substances had been discussed by biologists for years. And Lady Fleming has told us in a paper which, I believe, is not yet published, how greatly her late husband was interested in these questions, and in the medical possibilities of such substances.

Thus Fleming was not a passive observer of an accident. So far as it was an accident, it was one that happened to a well-prepared mind – a mind aware of the possible significance and desirability of 'accidents' of this kind. But an innocent reader of Fleming's paper would hardly suspect it. And this is the result of the traditional inductive style which, in its turn, comes of a mistaken view of scientific objectivity.

Now the practical proposal I wish to make is this. We should, as a matter of course, give the widest freedom to scientists to write papers as they think fit. But we could nevertheless encourage a new style, a style totally different from the traditional one.

A paper written in this new style might be in the following form:

It would start with a brief but clear statement of the problem situation as it stood before the research was started, and with a brief survey of the position reached so far in the discussion. It would then proceed to state briefly any hunch or conjecture related to the problem which may have motivated the research, and say which hypotheses the research hoped to test. Next it would outline the experimental arrangements, adding, if possible, reasons for choosing them, and the results. And it would conclude with a summary which would state whether any tests had been successful, whether the problem situation had changed in the opinion of the author, and if so, in what way. This part would also contain new hypotheses, if any, and perhaps some comment on how they could be tested.

Papers have been written in this style, some of them upon my suggestion. They were not all kindly received by the editors. But I believe that in the present situation of science in which high specialization is about to create an even higher Tower of Babel, the replacement of the inductive style by something like this new critical style is one of the few ways in which mutual interest and mutual contacts between the various fields of research can be preserved, or rather recreated. And I hope that the interest of the intelligent layman may also be rekindled in this way.

All this, of course, is merely a proposal open to discussion. But these matters ought to be discussed. For there does not seem to have been much discussion of questions like this for a long time – perhaps not even since Bacon, almost 400 years ago.

XV

I now come to the brief concluding part of my talk, entitled 'Responsibilities'.

Anybody who says anything today about the human or social responsibilities of scientists is expected, I am afraid, to say something about the bomb. So let me get the bomb out of the way first, for what I really wish to discuss has nothing to do with it.

Far be it from me to belittle the danger of nuclear warfare. The danger is terrible, as we all know, and the prospects of avoiding this kind of warfare are not as good as one could wish. This being so, we should try to make the best of a very unpleasant situation. It seems very likely that we shall have to live for a long time under the shadow of the bomb, and the only thing most of us can do, as far as I am able to see, is to accept the situation.

One of the things which we should avoid as far as possible is to become hysterical about it, and to proclaim loudly that this danger is the responsibility of us all.

There is very good reason for saying that road accidents are the responsibility of us all, because we are all users of roads, and we are all liable to make mistakes at times, as drivers or as pedestrians. But with the possible exception of a very small number of political or military leaders, we cannot do anything sensible about the danger of nuclear warfare.

In saying this I am taking a line which is rather the opposite of that which many worthy and well-informed people are taking. There has appeared, for example, quite recently a leading article in that interesting periodical, *The Bulletin of the Atomic Scientists*, which first developed a philosophical argument against fatalism and determinism, and went on to conclude that we are all responsible for what is going to happen – that the situation is a most urgent and desperate one, and that we should all do something about it as quickly as possible.

The author did not say what we should do. I suppose he thought that everybody should do his best according to his or her particular situation.

I think this author was wrong. I do not think it would be helpful if millions of citizens began to feel that they just had to do something about the bomb, and that they would be irresponsible, and fail in their duty as citizens, if they did not do something to prevent nuclear warfare. It seems to me possible even that an outbreak of this kind of feeling (which I personally would be inclined to describe as hysterical) might well add to the danger of nuclear attack.

A fact of life that we had better face is that sometimes we are involved in situations about which we should be ready to do whatever can be done, but about which we happen to be unable to do anything.

I do not wish to be dogmatic, and any practical suggestion or proposal should be most carefully discussed. This holds, of course, for the proposal called 'unilateral disarmament'. Yet even though I have always been a great admirer of Bertrand Russell as a philosopher, I feel that such proposals as unilateral disarmament have nothing whatever to recommend them. It seems strange to me that the propagators of unilateral disarmament never consider the possibility that if they were more successful in their propaganda, so that our determination to resist were seriously weakened, they might easily precipitate a nuclear attack. After all, there can be little doubt that the 18 years of uneasy nuclear peace enjoyed by us were very largely due to our readiness to fight. In other words, practical experience has shown that nuclear armament, dangerous as it is, may postpone the outbreak of nuclear warfare – perhaps for a sufficiently long time to lead to controlled disarmament. On the other hand, Hiroshima and Nagasaki have shown that if only one side in a conflict possesses atom bombs, it may well decide to use them in order to bring the conflict to an end (and to a speedy end, if possible, before the other side decides to build up – or to rebuild – an atomic arsenal).

Without having even the slightest inclination towards fatalism, I feel, like the famous 'man in the street', that those who cannot do anything about this danger should recognize this fact and learn to live with the danger as well as they can.

But I do think that, quite apart from the bomb, there are many sides to the present uneasy situation about which we can do something, and about which scientists, more especially, can do much, in purely peaceful ways.

Both our society and that of the Russians have a common background – the secularized religion of science. I mean the Baconian belief which grew during the Enlightenment that man may, through knowledge, liberate himself – that he may free his mind from prejudice and parochialism.

Like every great idea, this idea of self-emancipation through knowledge has, as we now know, its obvious dangers. Yet it is a very great idea. At any rate, we have embraced it. And though we can refine it, and develop it, we certainly cannot repudiate it now without condemning a large part of humanity to death by starvation.

Marxism calls itself a science. It is not a science, as I have tried to show elsewhere.[14] Yet in calling itself a science, it pays homage to science and to the idea of self-emancipation through knowledge. Much of its seductive power is connected with this fact.

At any rate Marxism, even though it has produced a ruthless dictatorship and an arrogant contempt for freedom and for individual human beings, is committed, like ourselves, to the idea of self-emancipation through knowledge – through the growth of science.

Thus there is a field of peaceful competition here, and one in which we can hardly fail, if we enter it whole-heartedly. The most important task for scientists in this competition is, of course, to do good work in their own particular fields. The second task is to shun the danger of narrow specialization: a scientist who does not take a burning interest in other fields of science excludes himself from participation in that self-liberation through knowledge which is the cultural task of science. A third task is to help others to understand his field and his work, and this is not easy. It means reducing scientific jargon to the minimum – that jargon in which many of us take pride, almost as if it were a coat of arms or an Oxford accent. Pride of this kind is understandable. But it is a mistake. It should be our pride to teach ourselves as well as we can always to speak as simply and clearly and unpretentiously as possible, and to avoid like the plague the suggestion that we are in the possession of knowledge which is too deep to be clearly and simply expressed.

This, I believe, is one of the greatest and most urgent social responsibilities of scientists. It may be the greatest. For this task is closely linked with the survival of an open society and of democracy.

An open society (that is, a society based on the idea of not merely tolerating dissenting opinions but respecting them) and a democracy (that is, a form of government devoted to the protection of an open society) cannot flourish if science becomes the exclusive possession of a closed set of specialists.

I believe that the habit of always stating as clearly as possible our problem, as well as the present state of the discussion of the problem, would do much to help towards the important task of making science – that is to say, scientific ideas – better and more widely understood.

NOTES

1 The quotation is taken from 'The Third Day'. The translation is my own. Cp. Stillman Drake's translation, *Dialogue Concerning the Two Chief World Systems*, University of California Press, Berkeley and Los Angeles, 1953, pp. 327f.

2 See my *Logic of Scientific Discovery* and chapter 1 of my *Conjectures and Refutations*.

3 See, for example, J.C. Eccles, *The Physiology of Nerve Cells*, Johns Hopkins University Press, Baltimore and Oxford, 1957, pp. 182–4.

4 When writing this I ought to have remembered the strange period from about 1929 or 1930 to 1932 or 1933, now easily forgotten, when the same feeling as described by Planck emerged again, though only for a short time, among some leading physicists. It is described by C.P. Snow in *The Search* where a Cambridge physicist whom he describes as 'one of the greatest mathematical physicists' and as 'Newton's successor' is made to say: 'In a sense, physics and chemistry are finished sciences.' (Penguin edition, London, 1965, p. 162. See also p. 88 for suggestions about the identity of the physicist.) A somewhat similar attitude may be discerned in R.A. Millikan's *Time, Matter and Values*, University of North Carolina Press, Chapel Hill, 1932, p. 46. The 'finished science' of those days was *the electrical theory of matter*, that is, *the theory of protons and electrons*: the structure of matter was to be explained by electrical forces (and even gravitation might well in the end be reduced to electricity). This theory of matter, which completely dominated the first third of the century, has slowly and almost silently disappeared –

certainly without causing anything like a violent, or even a conscious, revolution. (It should be remembered, in this context, that at that time quantum mechanics was the theory of electrons and of their behaviour in electrical fields, especially in the electrostatic fields of positively charged nuclei.)

5 What follows here, up to and including the first three paragraphs of section IX, is taken over, with very little change, from my Herbert Spencer Lecture of 1961. When I was giving the present lecture, I did not intend to publish the Spencer Lecture. But I have now published it as chapter 7 of my *Objective Knowledge*.

6 Albert Einstein, *On the Methods of Theoretical Physics*, Clarendon Press, Oxford, 1933. (Also in Albert Einstein, *The World as I See It*, translated by Alan Harris, Watts, London, 1940.)

7 *More Letters of Charles Darwin*, edited by Francis Darwin and A.C. Seward, Appleton, New York, 1903, volume I, p. 195. Darwin's comment ends with the words (which I admit weaken it as a support of my thesis) 'if it is to be of any service!'

8 Cp. for example the so-called 'Transportation Problem of Linear Programming'. See S. Vajda, *An Introduction to Linear Programming and the Theory of Games*, Methuen, London, 1960.

9 Cp. G. Polya, *How to Solve It*, Princeton University Press, Princeton, NJ, 1948.

10 The criticism by which we try to discover the weak spots of our theories leads to new problems. And by the distance between our original problems and these new problems we can gauge the progress made. Cp. my *Conjectures and Refutations*, p. 313.

11 See my *Logic of Scientific Discovery* and my *Conjectures and Refutations*.

12 I may perhaps note, in this context, that the much suspected term 'truth' in the sense of 'correspondence to the facts' has been rehabilitated (and shown to be innocuous) by Alfred Tarski, and that, using Tarski's theories, I have tried to do the same service to the terms 'better approximation to the truth' and of course 'less good approximation to the truth'. (See chapter 10 and the addenda of my *Conjectures and Refutations*.)

13 The *explanatory power* of a theory is discussed in my *Logic of Scientific Discovery*, as are some relevant meanings of the term 'simplicity' as applied to theories. More recently I have found it enlightening to interpret the simplicity of a theory as something that must be related to the *problems* which the theory is supposed to solve.

14 See my *Open Society* and my *Poverty of Historicism*.

5

PHILOSOPHY AND PHYSICS

The influence on theoretical and
experimental physics of some
metaphysical speculations on the
structure of matter

The following remarks are intended as an illustration of the
important thesis that science is capable of solving philosophical
problems and that modern science, at any rate, has something
important to say to the philosopher about some of the classical
problems of philosophy – especially about the old problem of
matter. I intend to discuss certain aspects of the problem of matter
since Descartes. And I intend to point out the interesting fact that
some of these problems were solved, in collaboration, by speculative
philosophers, such as Descartes, Leibniz, and Kant, who all helped
by proposing important though tentative solutions and thus prepared the way for the work of experimental scientists and theorists
of physics such as Faraday, Maxwell, Einstein, de Broglie, and
Schrödinger.

The history of the problem of matter has been sketched before,
notably by Maxwell.[1] But though Maxwell gives an outline of the
history of the relevant philosophical and physical ideas, he does not
give the history of the problem situation, and of how it changed
under the impact of the attempted solutions. It is this lacuna that I
am now trying to fill.[2]

Descartes based the whole of his physics upon an essentialist[3] or

Revised version of a paper first published in *Atti del XII Congresso Internazionale
di Filosofia, Venezia, 1958*, volume 2, Florence, 1960, pp. 367–74. This paper was
written as a critical response to the report delivered to the same Congress by my old
friend Philipp Frank, successor to Einstein's chair of theoretical physics in Prague,
and a leading member of the Vienna Circle of logical positivists.

Aristotelian definition of body or matter: a body is, in its essence or substance, extended; and matter is, in its essence or substance, *extension*. (Thus matter is extended substance, as opposed to mind which, as thinking or experiencing substance, is in essence *intensity*.) Since body or matter is identical with extension, all extension, all space, is body or matter: the world is full, there is no void. This is Parmenides' theory, as Descartes understood it. But while Parmenides concluded that there can be no motion in a full world, Descartes accepted a suggestion from Plato's *Timaeus* according to which motion is possible in a full world, as it is in a bucket of water: things may move in a full world like tea-leaves in a teacup.[4]

In this Cartesian world, all causation is action by contact: it is push. In a *plenum*, an extended body can move only by pushing other bodies. All physical change must be explicable in terms of mechanisms that work like cogwheels in clocks, or like vortices: the various moving parts *pushing* one another along. Push is the principle of mechanical explanation, of causation. There can be no action at a distance. (Newton himself sometimes felt that action at a distance was absurd, and at other times that it was supernatural.)

This Cartesian system of speculative mechanics was criticized by Leibniz on purely speculative grounds. Leibniz accepted the fundamental Cartesian equation, *body=extension*. But while Descartes believed that his equation was irreducible, self-evident, 'clear and distinct', and that it entailed the principle of action by push, Leibniz questioned all this: if a body pushes another body along instead of penetrating it, then this can only be because they both resist penetration. So this resistance must be *essential* to matter (or to bodies) – for it enables matter or body to fill space, and thus be extended in the Cartesian sense.

According to Leibniz, we must explain this resistance as due to *forces*: a body has 'a force and an inclination, as it were, to retain its state, and ... resist the cause of change'.[5] There are forces that resist interpenetration: repelling or repulsive forces. Thus body, or matter, in Leibniz's theory, is *space filled by repulsive forces*.

This is a programme for a theory that explains both the Cartesian essential property of body – that is *extension* – and the Cartesian principle of causation by push.

Since body or matter or physical extension is to be explained as due to forces filling space, Leibniz's theory is a theory of the *structure* of matter, like atomism. But Leibniz rejected the theory of atoms (which he had believed in when he was very young). For atoms, at this time, were nothing but very small bodies, very small bits of matter, very small *extensions*. So the *problem* of their extension and impenetrability was precisely the same for atoms as for larger bodies: extended atoms could not help to explain extension, the most fundamental of all the properties of matter.

In what sense, however, can a part of space be said to be 'filled' by repulsive forces? Leibniz conceives these forces as emanating from unextended points and thus as located ('located' only in the sense that they emanated from them) in unextended points, the *monads*: they are central forces whose centres are these unextended points. (Being an *intensity attached to a point*, a force may be compared, say, to the slope (or 'inclination') of a curve at a point, that is, to a 'differential': forces cannot be said to be 'extended' any more than differentials, though their intensities may of course be measurable and expressible by numbers; and being unextended intensities, forces cannot be 'material' in the Cartesian sense). Thus an *extended piece of space* – a body in the geometrical sense (a volume integral) – may be said to be 'filled' by these forces in the sense in which it is 'filled' by the geometrical points or 'monads' that fall within it.

For Leibniz, as for Descartes, there can be no void – empty space would be space free of repulsive forces, and since it would not resist occupation, it would at once be occupied by matter. One might describe this theory of the diplomat Leibniz as a political *theory of matter*: bodies, like sovereign states, have borders or limits which must be defended by repulsive forces; and a physical vacuum, like a political power vacuum, cannot exist because it would at once be occupied by the surrounding bodies (or states). Thus we might say that there is a general pressure in the world resulting from the action of the repulsive forces, and that even where there is no movement there must be a dynamic equilibrium due to the equality of the forces present. While Descartes could not explain an equilibrium except as mere absence of movement, Leibniz can explain an equilibrium – and also the absence of movement – as dynamically maintained by equal and opposite forces (whose intensity may be very great).

So much for the doctrine of point-atomism (or of monads) which grew out of Leibniz's criticism of the Cartesian theory of matter. His doctrine is clearly metaphysical. And it gives rise to a metaphysical research programme: that of explaining the (Cartesian) extension of bodies with the help of a theory of forces.

This programme was carried out in detail by Boscovitch (who was anticipated by Kant).[6] The contributions of Kant and Boscovitch will perhaps be better appreciated if I first say a few words about atomism in its relation to Newton's dynamics.

The no-vacuum theory of the Eleatic-Platonic school and of Descartes and Leibniz has one great difficulty – the problem of the compressibility and elasticity of bodies. Yet Democritus' theory of 'atoms and the void' (this was the password of atomism) had been designed, very largely, to meet precisely this difficulty. The void between atoms, the porosity of matter, was to explain the possibility of compressing and expanding it. But Newton's (and Leibniz's) dynamics created a new and grave difficulty for the atomistic theory of elasticity. Atoms were small bits of matter, and if compressibility and elasticity were to be explained by the movement of atoms in the void, atoms could not, in their turn, be compressible or elastic. They had to be absolutely incompressible, absolutely hard, absolutely inelastic. (This is how Newton conceived of them.) On the other hand, there could be no push, no action by contact, between *inelastic bodies* according to any dynamic theory which – like that of Newton or Leibniz – explained forces as proportional to *accelerations* (in a finite unit of time). For a push given by an absolutely inelastic body to another such body would have to be *instantaneous* (*and* of finite magnitude in the instant), and an instantaneous finite acceleration would be an *infinite* acceleration (in the unit of time), involving infinitely large forces.[7]

Thus only an *elastic* push can be explained by finite forces. And this means that we have to assume that all push is elastic. Now if we wish to explain elastic push within a theory of inelastic atoms, *we have to give up action by contact altogether*. In its place we have to put *short-distance repulsive forces between atoms*, or, as it might be called, action at short distance, or action in the vicinity: the atoms must repel each other with forces which rapidly increase with decreasing distance (and which would become infinite when the distance became zero).

In this way we are compelled, by the internal logic of the dynamic theory of matter, to admit central repulsive forces into mechanics. But if we admit these, then one of the two fundamental assumptions of atomism – the assumption that atoms are small extended bodies – becomes redundant. And since we have to replace the atoms by Leibnizian centres of repulsive forces, we might just as well replace them by Leibnizian unextended points: we can identify the atoms with Leibnizian monads which are nothing but repulsive forces. It seems, however, that we must retain the other fundamental assumption of atomism: the void. Since the repulsive forces tend towards infinity if the distance between the atoms or monads tends to zero, it is clear that there have to be *finite distances* between monads: matter consists of a void in which there are discrete centres of force.

The steps here described were taken by Kant and by Boscovitch. They may be said to give a synthesis of the ideas of Leibniz, of Democritus, and of Newton. The theory, like that of Leibniz, is a theory of the *structure of matter*, and thus a *theory of matter*. Extended matter is here *explained*, and by something that is not matter: by unextended entities such as forces and monads, the unextended points from which the forces emanate. The Cartesian extension of matter, more especially, is explained by this theory in a highly satisfactory way. Indeed, the theory does more: it is a *dynamic theory of extension* which explains not only equilibrium extension – the extension of a body when all the forces, attractive and repulsive, are in equilibrium – but also extension changing under external pressure, or impact, or push.[8]

There is another development, almost equally important, of the Cartesian theory of matter and of Leibniz's programme of a dynamic explanation of matter: while the Kant–Boscovitch theory anticipates in rough outline the modern theory of extended matter as composed of elementary particles invested with repulsive and attractive forces, this second development is the direct forerunner of the Faraday–Maxwell theory of fields.

The decisive step in this development is to be found in Kant's *Metaphysical Foundations of Natural Science* in which he repudiates[9] the doctrine that matter is discontinuous, which he had himself upheld in his *Monadology*. He now replaces this doctrine by that of the *dynamic continuity of matter*. His argument may be put as follows.

The presence of (extended) matter in a certain region of space is a phenomenon consisting of the presence of repulsive forces in that region, forces capable of stopping penetration (or forces which are at least equal to the attractive forces plus the pressure at that place). It is, accordingly, absurd to assume that matter consists of monads from which repulsive forces radiate. For matter would be present at places where these monads are not present, but where the forces emanating from them are strong enough to stop other matter. Moreover, it would be present for the same reason *at any point between* any two monads belonging to (and allegedly constituting) the piece of matter in question.

Now whatever the merits of this argument may be,[10] there is at any rate great merit in the proposal to try out (and perhaps make more specific) the vague idea of a continuous (and elastic) something – of an entity consisting in the presence of forces. For this is simply the idea of a continuous field of forces in the guise of the idea of continuous matter. It seems to me an interesting fact that this second dynamic explanation of (Cartesian) extended matter and of elasticity was mathematically developed by Poisson and Cauchy, and that the mathematical form of Faraday's idea of a field of forces, due to Maxwell, might be described as a development of Cauchy's form of Kant's continuity theory.

Thus the theory of Boscovitch and the two theories of Kant may be described as the two most important attempts to carry further Leibniz's programme for a dynamic theory that explains Cartesian extended matter. They may be described as the joint ancestors of all modern theories of the structure of matter; the theories of Faraday and Maxwell, of Einstein, de Broglie and Schrödinger, and also of the 'dualism of matter and field'. (This dualism, if seen in this light, is perhaps not so deep as it may appear to those who, in thinking of matter, cannot get away from a crude Cartesian and non-dynamical model.) It may be mentioned that another important influence deriving from the Cartesian tradition – and from the Kantian tradition via Helmholtz – was the idea of explaining atoms as vortices of the ether – an idea that led to Lord Kelvin's and to J.J. Thomson's models of the atom. Its experimental refutation by Rutherford marks the beginning of what may be described as modern atomic theory.

One of the most interesting aspects of the development which I have sketched is its purely speculative character, together with the fact that these metaphysical speculations proved *susceptible to criticism*: that they could be critically discussed. This discussion was inspired by the wish to understand the world, and by the hope, the conviction, that the human intellect could at least make the attempt to understand it, and could perhaps get somewhere. And an experimental refutation of a speculative solution to one of its problems led to its turning into nuclear science.

Positivism, from Berkeley to Mach, was always opposed to these speculations. And it is most interesting to see that Mach still upheld the view that there could be no physical theory of matter (which for him was nothing but a metaphysical 'substance' and as such redundant if not meaningless) at a time (after 1905) when the metaphysical theory of the atomic structure of matter had turned into a testable physical theory as a result of Einstein's theory of Brownian motion. It is perhaps somewhat ironical, and certainly more interesting, that these views of Mach's reached the peak of their influence when the atomic theory was no longer seriously doubted by anybody, and that it is still most influential among the leaders of atomic physics, especially Bohr, Heisenberg, and Pauli.[11]

Yet the wonderful theories of these great physicists are the result of attempts to understand the structure of the physical world, and to criticize the outcome of these attempts. Thus their theories may well be contrasted with what they, and other positivists, try to tell us today: that we cannot, in principle, hope ever to understand anything about the structure of matter; that the theory of matter must forever remain the private affair of the expert, the specialist – a mystery shrouded in technicalities, in mathematical techniques, and in 'semantics'; that science is nothing but an instrument, void of any philosophical or theoretical interest, and only of 'technological' or 'pragmatic' or 'operational' significance. I do not believe a word of this 'post-rationalist' teaching. Nothing could be more impressive than the progress made in our attempts – and especially the attempts of these great physicists – to understand the physical world. No doubt, we shall modify and even discard our theories many times. But it seems that we have at last found a way towards the understanding of the physical world.

NOTES

1 See Maxwell's masterly article *Atom* in the ninth edition of the *Encyclopedia Britannica*.
2 For many years I have been in the habit of giving an outline of the story (which begins with Hesiod) in my lectures.
3 I have criticized (Aristotelian) *essentialism* and also the *essential theory of definitions* in my books *The Open Society and Its Enemies* and *The Poverty of Historicism*; see the Indices, under 'essence' and 'essentialism'.
4 Descartes, *Principia Philosophiae*, Elzevir, Amsterdam, 1644, part II, point 33f. By asserting the infinite divisibility of matter, Descartes prepared the way for Leibniz's non-extended monads. (Monad=point. A point is unextended and *therefore* immaterial.) In II, 36 Descartes asserts the conservation of the 'quantity of motion' (*quantitas motūs*): God Himself 'who in the beginning created matter together with motion and rest, conserves in its totality as much of motion and rest as he originally put into it'. Note that this 'quantity of motion' is *neither*

our 'momentum', which has a definite *direction* and which is indeed conserved, *nor* our 'angular momentum' but, rather, mass times the (non-vectorial) *amount* of velocity which, as Leibniz showed (*Mathematische Schriften*, edited by C.I. Gerhardt, Weidmann, Berlin and Halle, 1849–63, volume VI, pp. 117ff.), is not conserved. (On the other hand, 'force' – which Leibniz thought was conserved – is not conserved either, not even *vis viva* ($mv^2/2$), that is, kinetic energy. The fact is that both Descartes and Leibniz had an intuitive idea of the conservation laws, and though Leibniz came nearer to the truth than Descartes, neither got very near.)

5 Leibniz, *Philosophische Schriften*, edited by C.I. Gerhardt, Weidmann, Berlin, 1875–90, volume II, p. 170, lines 27f. J.W.N. Watkins has developed this argument in some detail, showing that for these ideas Leibniz was essentially dependent on Hobbes, whose term '*conatus*' (translated into English by 'endeavour') Leibniz adopted, and which he identified with force. See J.W.N. Watkins, *Hobbes's System of Ideas*, Hutchinson, London, 1965, pp. 122–32; 2nd edition, 1973, pp. 85–94.

6 Boscovitch's *Philosophiae Naturalis Theoria Redacta ad Unicam Legem Virium in Natura Existentium* was first published in 1758 in Vienna (the second, improved, edition was translated by J.M. Child as *Theory of Natural Philosophy* and published in London in 1922), Kant's *Metaphysicae cum geometria iunctae usus in philosophia naturali, cuius · specimen I. continet Monadologiam physicam* (referred to in English as '*Monadology*') in 1756 in Königsberg. Thirty years later Kant repudiated part of his *Monadology* in his *Metaphysische Anfangsgründe der Naturwissenschaft*, published in Riga in 1786 (translated by James W. Ellington as *Metaphysical Foundations of Natural Science*, Bobbs-Merrill, Indianapolis, 1970). Though the essential idea of Boscovitch's monadology is to be found in Kant (see Kant, proposition V for the finite number of *discrete* monads present in finite bodies, and proposition X for the central forces which are attractive over long distances and repulsive over short distances, and for Kant's explanation of extension), Kant's work is rather sketchy as compared with Boscovitch's. (Added 1973.) Restrictions on the size of this paper when it was originally presented prevented me from discussing Faraday. While Boscovitch developed the Newtonian research programme which treated physical events as due to central forces (acting, however, at infinitely small distances – from one point to the next, as it were), it was Faraday's revolutionary innovation that he broke with the dogma of central forces. Although Maxwell, with his models, still hoped, like Ampère, to reduce non-central forces to central forces, his theory in effect also broke with that dogma. It thereby achieved a generalization which, I suggest, inaugurated modern field theory, and so led to special and general relativity.

7 The argument is clearly stated by Kant in his *Metaphysical Foundations of Natural Science*, 'General Observation on Mechanics', pp. 115–17. See also his *Monadology*, proposition XIII, and his '*Neuer Lehrbegriff der Bewegung und Ruhe*', 1758 (the section on the Principle of Continuity). Similar arguments may be found in Leibniz (see his

Mathematische Schriften, volume II, p. 145), who says that it seems that only elasticity 'makes bodies rebound'. The best statement of the argument is to be found in the work of Boscovitch.

8 It is important to realize that Boscovitch's forces are not to be identified with Newtonian forces: they are *not* equal to acceleration multiplied by mass, but equal to acceleration multiplied by a pure number (the number of monads). This point has been clarified by L.L. Whyte (in a very interesting note in *Nature*, 179, 1957, pp. 284f.). Whyte stresses the 'kinematic' aspects of Boscovitch's theory (as opposed to its 'dynamical' aspects, in the sense of Newton's dynamics). It seems to me that Whyte's reply to Maxwell is correct. I may perhaps express this by saying that Boscovitch not only gives a theory of extension and gravity but also of Newtonian inertial mass. On the other hand, although Boscovitch's forces are, as Whyte rightly stresses, from a formal or dimensional point of view *accelerations*, they are, from a physical and a metaphysical point of view, *forces*, very much like Newton's: they are dispositions, existing in their own right; they are the *causes* that *determine* accelerations. Kant, on the other hand, thinks in purely Newtonian terms, and attributes inertia to his monads; see his *Monadology*, proposition XI.

9 See the second chapter, theorem 4, especially the first paragraph of note 1, and note 2. Kant's repudiation is the result of that doctrine which he calls (in the *Critique of Pure Reason*) his 'transcendental idealism': he rejects the monadology as a doctrine of *the spatial structure* of things in themselves. (This way of speaking would be for him a mixture of spheres – something like a 'category mistake'.)

10 Like all such proofs Kant's proof is invalid, even in the form here given, which is an attempt to improve a little on Kant's own version. Kant illicitly identifies 'moving', in the sense of a moving (repelling) force, and 'movable'; cp. the penultimate paragraph of his note 1 to theorem 4. The ambiguity is bad, but it brings out the fact that he *wishes* to identify the presence of a moving force with that of movable matter. The logical situation is, in brief, as follows. In this post-critical work, Kant's transcendental idealism is used to remove – by a valid argument, incidentally – his original *objections* to the doctrine of continuous matter. But he now thinks, mistakenly, that he can *prove* continuity – by an argument which, though invalid, is interesting and important because it compelled him to push his dynamics to its very limits (and far beyond those limits which he anticipated in his definitions).

11 Niels Bohr, Werner Heisenberg, and Wolfgang Pauli were all alive when this was written.

6

THE MORAL
RESPONSIBILITY OF
THE SCIENTIST

The topic I am going to discuss was not of my choosing, but was suggested by the organizers of this conference. I say this because I do not think that I can make any significant contribution to the solution of the grave problems involved. I nevertheless accepted the invitation to speak about it because I believe that in this respect we are all more or less in the same boat. I take it that our topic, 'the moral responsibility of the scientist', is a kind of euphemism for the issue of nuclear and biological warfare. But I shall try to approach our topic with some wider issues in mind.

One may say that the problem has lately become more general, due to the fact that lately all science, and indeed all learning, has tended to become potentially applicable. Formerly, the pure scientist or the pure scholar had only one responsibility beyond those which everyone else has – that is, to search for truth. He had to further the growth of his subject as well as he could. For all I know, Maxwell had little reason to worry about the possible applications of his equations. And perhaps even Hertz did not worry about Hertzian waves. This happy situation belongs to the past. Today not only all pure science may become applied science, but even all pure scholarship.

In applied science the problem of moral responsibility is a very old one, and like many other problems, it was first posed by the

This paper is a revised version of a brief address read, on 3 September 1968, to a special session, 'Science and Ethics: The Moral Responsibility of the Scientist', at the International Congress of Philosophy, held in Vienna. A version of this paper was published in *Encounter* in March, 1969, and a revised version in the *Bulletin of Peace Proposals*, Oslo, 1971. The paper has been further revised for the present publication. I am greatly indebted to my friend Ernst Gombrich for his help in the preparation of the original address.

Greeks. I have in mind the Hippocratic Oath,[1] a marvellous document even though some of its main ideas may be in need of renewed scrutiny. I myself took an oath which no doubt historically derives from the Hippocratic Oath when I graduated from the University of Vienna. One of the most interesting points about the Hippocratic Oath is that it was not a graduation oath but an oath to be taken by the apprentice to the medical profession. Essentially, it was taken at the beginning of the student's first initiation into applied science.

The Oath consisted in the main of three parts. First, the apprentice undertook to recognize his deep personal obligation to his teacher. By implication, this obligation is considered to be mutual. Secondly, the apprentice promised to carry on the tradition of his art, and to preserve its high standards, dominated by the idea of the sanctity of life, and to hand on these standards to his own students. Thirdly, he promised that to whatever house he would go, he would enter it only in order to help the suffering, and that he would preserve silence about whatever may become known to him in the course of his practice.

I have stressed the fact that the Hippocratic Oath is an apprentice's oath because in many discussions of our topic the situation of the apprentice, that is of the student, is not sufficiently considered. However, prospective students are worried about the moral responsibility which they will have to carry once they become creative scientists, and I feel it may be of considerable help if they have an opportunity to discuss these issues at the beginning of their studies. Ethical discussions, unfortunately, tend to become somewhat abstract, and I suggest that we take the opportunity to make the issues more concrete. My proposal would be that we try to hammer out a modern form of an undertaking analogous to the Hippocratic Oath, in cooperation with our students.

It is obvious that no such formula should be *imposed* upon the students. If they object, they would thereby show a most welcome interest and they should be asked to offer an alternative approach or give reasons for objecting. The main purpose would be to draw their attention to the significance of the issues and so to keep the discussions going.

May I perhaps give an indication of the sort of thing that I have in mind as a possible starting-point for discussion.

I should propose that the order of the Hippocratic Oath be inverted, according to the significance of the various points. Thus my own points *1*, *2*, and *3* below would roughly correspond to the parts *2*, *1*, and *3* of the Hippocratic Oath as summarized above. I would also suggest that the main issues of the Oath might be generalized, somewhat along the following lines.

1. Professional Responsibility. The first duty of every serious student is to further the growth of knowledge by participating in the search for truth – or in the search for better approximations to the truth. Of course, every student is fallible, as are even the greatest masters: everybody is bound to make mistakes – even the greatest thinkers. Although this fact should encourage us not to take our mistakes over-seriously, we must resist the temptation to look upon our mistakes leniently: the establishment of high standards by which to judge our work, and the duty to constantly raise these standards by hard work, are both indispensable. At the same time, we must constantly remind ourselves (especially in connection with the application of science) of the finitude and fallibility of our knowledge, and of the infinity of our ignorance.

2. The Student. He belongs to a tradition and to a community, and he owes respect to all who have contributed, or are contributing, to the search for truth. He also owes loyalty to all his teachers who freely and generously share their knowledge and enthusiasm with him. At the same time, he has a duty to be critical towards others, including his teachers and colleagues, and especially towards himself. And most important, he has the duty to beware of intellectual arrogance, and to try not to succumb to intellectual fashions.

3. The Overriding Loyalty. This he owes neither to his teacher nor to his colleagues, but to mankind – just as the physician owes his overriding loyalty to his patients. The student must constantly be aware of the fact that every kind of study may produce results which may affect the lives of many people, and he must constantly try to foresee, and guard against, any possible danger, or possible misuse of his results, even if he does not wish to have his results applied.

This is a very tentative restatement of the Hippocratic Oath, at best a proposal for renewed discussion, and I must stress that all this is merely peripheral to our topic. But I have started with this practical proposal because I believe both in traditions and in the need for their continuous critical revision. One of the few things we can do about our main issue is to try to keep alive, in all scientists, the consciousness of their responsibility.

In this context, one point should be mentioned, a point which I think may be connected with the present [1968] crisis of the universities. It is this. More and more technicians are needed, and as a consequence, more and more Ph.D. students are trained only as technicians. Often they are trained only in measuring techniques. And they are not even told what more fundamental problems are to be solved by the measurements they are doing for their doctor's thesis. I regard this situation as inexcusable and irresponsible. I see in it a kind of breach of the Hippocratic Oath on the side of the academic teacher. For his task is to initiate the student into a tradition, and to explain to him the new great problems which arise through the growth of knowledge and which in their turn inspire and motivate all further growth.

I know, of course, that even the beautiful tradition of the Hippocratic Oath can be misused, and that it has been misused or misunderstood by interpreting it as establishing a special ethical obligation towards one's professional colleagues. It has, in other words, been interpreted as a kind of guild morality. It is precisely the serious discussion of issues like the gulf between *ethics* and *etiquette* ('professional ethics') which, I hope, may lead us to a much needed advance in our moral awareness. My hopes are modest: I do not think that by such discussions any of the great problems with which we are faced can be solved. But discussions centring on a revision of the Hippocratic Oath may lead to reflection on such fundamental moral problems as the priority of the alleviation of suffering.

Many years ago I proposed that the agenda for public policy should consist, in the first place, of finding *ways and means of avoiding suffering*, so far as it is avoidable. Contrasting this with the utilitarian principle of maximizing happiness, I proposed that, in the main, happiness should be, and that it can only be, left to private initiative, while the alleviation of avoidable suffering is a problem of public policy. I also indicated that at least some utilitarians, when speaking of the maximization of happiness, may have had in mind the minimization of misery.

Of course, I never suggested that the minimization of suffering should be made the highest moral principle. In fact I do not believe in the existence of such a thing as a single highest universally valid moral principle. On the contrary, I suggested that in matters of public policy we have to constantly reconsider our priorities, and that in drawing up a list of priorities avoidable suffering rather than happiness should be our main guide. Perhaps not forever: there may come a time when the alleviation of avoidable suffering will be less important than it is today.

Today the avoidance of war is, I should say by general consent, the overriding problem of public policy. There is no doubt in my mind that we all, whether as scientists, scholars, citizens, or mere human beings, should do everything we can to help to end war. It is part of this effort that we must try to make clear to everybody what war means, not only in terms of death and destruction, but also in terms of moral degradation. In this context it should be stated very clearly that one of the most disturbing aspects of recent events is the cult of violence. We all know that one of the most horrible aspects of our entertainment industry is the constant propaganda for violence, from allegedly harmless Westerns and crime stories to displays of cruelty pure and simple. It is tragic to see that this propaganda has had its effects even on genuine artists and scientists, and unfortunately also on our students (as the cult of revolutionary violence shows).

It is, however, my conviction that neither the First nor the Second World War, nor the present tragedy of Vietnam, can be explained in terms of human aggressiveness. At least today the main danger of war comes from the need to resist aggression, and from the fear of aggression. These, combined with muddleheadedness and lack of intellectual flexibility, and perhaps megalomania, tend to become the main sources of danger in the presence of the tremendous means of destruction which are at our disposal.

Only the problem of avoiding tyranny, the danger of losing our freedom (a loss which, in its turn, would ultimately lead to war), can compete in urgency with the problem of avoiding war – a competition which sometimes may make our decisions difficult.

Some people have thought that it is therefore the moral obligation of the scientist to withdraw from all military work, and to propagate disarmament at any price, including unilateral disarmament. I think that the situation is by no means as simple as that. We cannot shut our eyes to the fact that atomic war has so far been prevented by the danger of mutual destruction. So far, the deterrent has been successful in deterring. This is why I do not believe that we should support unilateral disarmament. The fact that Japan did not have atomic arms did not prevent us from using them. I do not think that this happened because we are morally inferior to our competitors in the armament race. The question whether we should have ever dropped the bomb on Japan is a very difficult one. The scientists who were in favour of its use were, I am sure, highly responsible people. Where I think they were wrong is that they did not insist

125

that the bomb should, in spite of the greater risk involved, be dropped on a *purely* military target, such as a concentration of warships – if it were to be dropped at all. (Such concentrations did exist at the time.) However, we should realize that decisions like these are frightful. It is all too easy to talk about such matters, but terrible to be involved, and to have to make up one's mind as to which decision would ultimately lead to a lesser amount of suffering. Nor must we forget that the politicians who were responsible for the ultimate decision were acting as trustees for those who elected them. This may be a reason for you or me not to become a politician, but it should not be a reason for you or me to glibly pass judgment on them.

One cannot back out altogether of the general involvement which is part of human life: everything has to be done to avoid a war and, if there is a war, to bring it to an end. This does not mean that there cannot be something like a just or defensive war. There is a world of difference between aggression and defence, even though it may not always be easy to decide who is the aggressor. Who believes that Switzerland or Sweden would nowadays wage an aggressive war? Who can believe for a moment that it was Serbia who attacked Austria in July 1914, or that it was Finland who attacked Russia on 30 November 1939, rather than the other way round? Or that Czechoslovakia has been threatening Russia?

A scientist who feels that his country is threatened by an attack cannot be blamed for working to defend his country. However, even a just war may get utterly out of hand, and it seems to me unlikely that there can be, or that there has ever been, a war without war crimes on both sides. Thus, once a war has started, the scientist, like any other citizen, is caught in a terrible moral difficulty, and nobody can give him advice or shoulder his responsibility.

One point can be made clear. It was the politicians and the law officers of the various Allied countries who staged the Nuremberg Trials which established the status of war crimes and thereby recognized that the conscience of every human being is the ultimate court of appeal with respect to the question whether a certain command is, or is not, to be resisted. It is impossible, without contradicting themselves, for these same politicians and law officers now to assert that it is the duty of the citizen – and of the scientist – not to ask the reason why and to obey any command. The freedom for which we must be prepared to fight is precisely the freedom to resist a command which we feel it would be criminal to obey. It is,

I believe, the inescapable duty of every loyal politician in a democracy to understand the terrible situation in which a scientist may find himself, and to champion the rights of the conscientious objector, whether he is a scientist or a soldier.

The trouble with the present legislation concerning conscientious objectors in the United States is that a man has to declare that he objects on conscientious grounds to *all* wars in order to plead conscientious objection. But there are people who would feel it their duty to fight for the United States, provided they can see that the war is waged for the defence of the United States, but who feel that they cannot conscientiously fight in Vietnam. Clearly such moral scruples should be respected as much as any that fall under the present definition of conscientious objection. Here, as always, I believe in the critical discussion of the issue involved, rather than in facile slogans from either side.

I discuss these very grave issues not because I believe in my ability to solve them or to say anything very new about them, but mainly because I feel that they should not be dodged. I am convinced, however, that the moral responsibility of the scientist is not confined to his responsibility in connection with war or armament.

The late Dr Robert Oppenheimer is reputed to have said: 'We scientists have been on the brink of presumptuousness in these years. We have known sin ...'. But this, again, is not a recent issue. When Bacon tried to make science attractive by saying that *knowledge is power*, he too was on the brink of presumptuousness. Not that he had much knowledge or much power, but he wanted knowledge because he wanted power – or at least he gave the impression that he did so.

I do not intend to philosophize about the wickedness of power in general, although my experience corroborates Lord Acton's saying that power tends to corrupt and that absolute power corrupts absolutely. As far as science is concerned, there is no doubt whatsoever in my mind that to look upon it as a means for increasing one's power is a sin against the Holy Ghost. The best antidote against this temptation is the awareness of how little we know and that the best of those little additions to our knowledge which we have achieved have shown their significance precisely by the fact that they opened up a whole new continent of our ignorance.

The social scientist has a particular responsibility here, because his studies more often than not concern the use and misuse of power pure and simple. I feel that one of the moral obligations of the social scientist which ought to be recognized is that, if he discovers tools of power, especially tools which may one day endanger freedom, he should not only warn the people of the dangers but devote himself to the discovery of effective counter-measures. I am confident that in fact most scientists, at least most creative scientists, value independent and critical thinking very highly. Most of them hate the very idea of a society manipulated by the technologists and by mass-communication. Most of them would agree that the dangers inherent in these technologies are comparable to those of totalitarianism. Yet although we built the atom bomb in order to combat totalitarianism, few of us regard it as our business to think of means to combat the dangers of mass-manipulation. And yet, there is no doubt in my mind that much should and could be done in this direction, without censorship or any similar restriction of freedom.

It could be questioned whether there is such a thing as a responsibility of the scientist which differs from that of any other citizen or any other human being. I think the answer is that everybody has a special responsibility in the field in which he has either special power or special knowledge. Thus, in the main, only scientists can gauge the implications of their discoveries. The layman, and thus the politician, does not know enough. This holds for such things as new chemicals for increasing the output of farming products as much as for new armaments. Just as, in former times, *noblesse oblige*, so now, as Professor Mercier has put it, *sagesse oblige*: it is the potential access to knowledge which creates the obligation. Only scientists can foresee the dangers, for example, of population increase, or of the increases in the consumption of oil products, or the dangers inherent in atomic waste, and thus even in atoms for peace. But do they know enough about it? Are they conscious of their responsibilities? Some of them are, but it seems to me that often they are not. Some, perhaps, are too busy. Others, perhaps, are too thoughtless. Somehow or other, the unintended repercussions of our heedless general technological advance seem to be nobody's business. The possibilities of applications seem to be intoxicating. Though many people have questioned whether techno-logical advance always makes us happier, few people make it their business to find out how much avoidable suffering is the unavoid-able, though unintended, consequence of technological advance.

The problem of the unintended consequences of our actions, consequences which are not only unintended but often very difficult to foresee, is the fundamental problem of the social scientist.

6. THE MORAL RESPONSIBILITY OF THE SCIENTIST

Since the natural scientist has become inextricably involved in the application of science he, too, should consider it one of his special responsibilities to foresee as far as possible the unintended consequences of his work and to draw attention, from the very beginning, to those which we should strive to avoid.

NOTE

1 See *Hippocrates*, with an English translation by W.H.S. Jones, volume I, Loeb Classical Library, William Heinemann, London/G.P. Putnam's Sons, New York, 1923, pp. 299–301.

7

A PLURALIST
APPROACH TO THE
PHILOSOPHY OF
HISTORY

I

What may be called the philosophy of history persistently turns round three big questions.

1   Is there a plot to history, and if so, what is it?
2   What is the use of history?
3   How are we to write history, or what is the method of history? (This also includes the 'problem of historical knowledge'.)

Answers to these three questions have been given, implicitly and explicitly, from the Bible and Homer down to our own day. And the answers have changed astonishingly little.

The oldest answer to our first question, given in the Bible and in Homer, is theistic. There *is* a plot to history. But it is only dimly discernible, because it results from the will of God, or of the gods. And though it is perhaps not completely unfathomable, it is not easy to fathom. At any rate, there is something secret hidden behind the surface of events. It has to do with reward and punishment, with a kind of divine balance of justice – though only by the most discerning can justice be seen to be done.

This balance, which if upset swings back like a pendulum, plays its part in Herodotus, who sees the eastward movement of the people in the Trojan War as an explanation of the subsequent swing

Based on a lecture delivered in Oxford on 3 November 1967. First published in *Roads to Freedom: Essays in Honour of Friedrich A. von Hayek*, edited by Erich Streissler, Gottfried Haberler, Friedrich A. Lutz, and Fritz Machlup, Routledge & Kegan Paul, London, 1969. This paper has been revised in the light of suggestions by Kims Collins, Morris Cranston, and Jeremy Shearmur.

back of the Persian Wars with their westward movement. Twenty-three centuries later we find precisely the same theory in Tolstoy's *War and Peace*: Napoleon's movement to the East into Russia is reciprocated by a movement of the Russian people to the West.

Admittedly, neither Herodotus nor Tolstoy offers what at first sight looks like a theistic theory. But the theistic background – a more or less suppressed theory of a divine balance of justice – is unmistakable. This, after all, is in keeping with the whole structure of European thought, which is fundamentally theological in origin, tenaciously clinging to its theological ground plan, despite anti-religious movements, despite the French revolution, and despite the rise of science. For the naturalistic revolution replaced the name 'God' with the name 'Nature', but left almost everything else unchanged.[1] Later, Hegel and Marx replaced the goddess Nature, in her turn, by the goddess History. So we get laws of History – powers, forces, tendencies, designs, and plans of History – and the omnipotence and omniscience of historical determinism. Sinners against God are replaced by 'criminals who vainly resist the march of History'. And we learn that not God but History will be our judge.[2]

It was the theory that there is a plot in history, whether theistic or anti-theistic, that I called by the name 'historicism'. My use of this term has been severely criticized by some people. Yet their criticisms seem to me without force, for they depend on the mistaken theory that names or terms matter. In fact the name 'historicism' is no more than a label that I introduced as a convenient way of talking about various connected theories that I was explaining and discussing. And I said as much when I introduced it (and, incidentally, I also explicitly pointed out that I was *not* discussing the doctrine of historical relativism, which I referred to as 'historism').[3]

My criticism of historicist theories has also been attacked for being out of date. There are no historicists left, it has been said. So why attack them?

Now it is quite true, especially more recently, that there have been few people who openly defend historicism. Even Marxists, and even followers of Professor Toynbee, have become in this respect less vocal, if not subdued. Nevertheless, I still feel as if I were almost drowning in a historicist flood. For we are constantly told that we are living in an atomic age, and in the space age – in the age of television, and in the age of mass communication. We are also constantly told about the age of specialization in which we live and, at the same time, about the age of abstract revolutionary art – which, incidentally, seems hardly to have changed since 1920 when practically all of its variations were exhibited in the *Bauhaus* in Weimar. It was then a revolutionary movement of protest against stagnation and conformity. But ever since it has stagnated, and it is still conforming to the pattern of a revolutionary movement of protest against stagnation and conformity.

I think that all this talk of movements and tendencies, ages and periods (and their 'spirits') signals the acceptance – tacit or otherwise – of theories that are clearly historicist in character: for example, theories of intrinsic historical progress or regress.[4] This comes out especially clearly when such ideas are used as if they were arguments for the acceptability of the thing in question (for example supersonic aircraft).

For the historicist, the 'Spirit of the Age' is an entity that explains largely, or at least partly, the actions and the sayings of the men living in that age. This approach seems to me quite mistaken. But this does not mean that there is no problem here. The spirit of the age must be demoted from an explanation to a social phenomenon that we have to explain. It is to be explained by the existence of overruling problems and problem situations, and by the interaction of individuals and their plans and aims – that is to say, in terms of situational logic.[5]

I am, however, alive to the dangers of stagnation, including the danger of letting my own ideas stagnate. And I will therefore not say anything further here against historicism.

On the contrary, I will ask here whether there is not, perhaps, a grain of truth in historicism, or more precisely in the historicist idea of a plot in history. In other words, I propose to take a fresh look, though only a *very* brief one, at my first question – Is there a plot to history? (or at least to human history) – and even to answer it by saying that, by and large, the answer seems to be 'yes'. (Though I wish to make it quite clear that I do not thereby weaken my criticism of historicism. I still regard historicism as badly mistaken.)

For since the invention of critical discussion and of writing, there has taken place something that may be described as the growth of knowledge. Knowledge, and its growth, have had a greater and greater influence on men's lives, both directly and by way of technological applications. It is, I suppose, only within the last two hundred years that the influence of the growth of scientific knowledge has become very obvious. But if we look back from our present vantage point, we can say, I think, not only that it is in our knowledge that we differ most clearly from other animals, but that the growth of knowledge – and of scientific knowledge – forms something like a plot of history. We can, I suggest, see the growth of our knowledge as a continuation of animal evolution (although by entirely new means). And thus, when we consider it from a biological point of view, we can see the growth of our knowledge not only as the main plot of human history, but perhaps also of the evolution of life.

This way of looking at our history is both obvious and extremely one-sided. For four hundred years ago the growth of scientific knowledge was not a historical fact but rather a dream – the dream of a highly dubious prophet, Francis Bacon. And Bacon's dream, after becoming a kind of programme for research, became, in its turn, a typical intellectual fashion. Nevertheless, I do think that, from our present standpoint, my suggestion is a reasonable one. But we should not, of course, forget that just as the survival of a species up to a certain point does not entitle us to say anything about its future survival, we cannot – and should not try to – derive predictions about the future from this 'plot' of human history.

Perhaps I overstated my case in saying that this way of looking at things is obvious. For not only do most professional historians ignore it, but they also seem to take very little interest in the history of science. As I remarked in my *Open Society*, the history of science is completely ignored in the first six volumes of Arnold Toynbee's huge *Study of History*. And in another well-known book, first published in 1938, by another very famous historian, the following strange remark can be found: '... the study of the material world had been revolutionized by Galileo's assertion that the world revolved round the sun'.

When I read this remark I was surprised. For after all, this particular revolution had originated, as everybody knows, with Copernicus, a full century before. I thought, for a moment, that the word 'assertion' might here mean reassertion. But the next sentence and some other passages showed me that this historian really had mistaken Galileo for Copernicus (or the other way round). For the next sentence begins with the unambiguous words 'Before Galileo's discovery' – words which refer back to the remark 'Galileo's assertion that the earth revolved round the sun'. And examples could be multiplied of historians' unfamiliarity with even the crudest outline of the history of science.

Incidentally, almost all creative scientists know a great deal about the history of their problems, and therefore about history. *They have to*: you cannot really understand a scientific theory without understanding its history.

It is to be hoped that historians will, in their turn, soon discover that they have to know something about science and its history. For one cannot really understand any recent history, least of all political or diplomatic history, without understanding something about science. In this they could learn from Churchill, in whose *The Second World War* an adequate treatment of the development of radar can be found.

But I do not regard it as my present task to complain further about the much-discussed gulf between the 'two cultures'. Let me, therefore, return to our first question, to the question of the plot of history.

I suggest that man has created a new kind of product or artefact that promises in time to work changes in our corner of the world as great as those worked by our predecessors, the oxygen-producing plants, or the island-building corals. These new products, which are so decidedly of our own making, are our myths, our ideas, and especially our scientific theories: our theories about the world we live in. Indeed we may look upon these myths, these ideas, and theories as some of the most characteristic products of human activity. Like tools, they are organs evolving outside our skins. They are exosomatic artefacts. Thus we may count among these character-istic products of man especially what is called 'human knowledge', where we take the word 'knowledge' in the objective or impersonal sense in which it may be said to be contained in a book, or stored in a library, or incorporated in a university curriculum.

When speaking here of human knowledge, I shall usually have this objective sense of the word 'knowledge' in mind. This allows us to think of knowledge produced by men as analogous to the honey produced by bees. The honey is made by bees, stored by bees, and consumed by bees. And the individual bee that consumes honey will not, in general, consume only the honey it has produced itself. Honey is also consumed by the drones, who have not produced any at all.

The same holds true, with slight variations, for theory-producing men. We, too, are not only producers but consumers of theories. And we have to consume other people's theories, and sometimes perhaps our own, if we are to produce more.

Thus the growth of human knowledge continues the evolution of other organisms. But because it is almost entirely exosomatic, and transmitted by tradition, it is something new, and characteristic of human history.

I have tried to give a very brief and somewhat sweeping answer to our first question, and this answer may appear to be a monistic rather than a pluralistic one: it may seem as if I were saying that the growth of knowledge, more particularly the history of science, is the heart of all history.

But this is not my intention. Admittedly, the life of all men is by now deeply affected by science. But the life of all men has also been deeply affected by religion (or religions). And the history of religion is at least as important as the history of science. Science itself is closely linked with religious myths: I am even inclined to say that there would have been no European science without Hesiod's *Theogony*.[6] What is more, while everybody is affected by the growth of knowledge, comparatively few people contribute to it. Religious beliefs, on the other hand, are shared and actively participated in by many people – as are some of the new religious movements and cults of the living gods of the cinema, the television, and the gramophone disc. Stars and starlets were gods for the Greeks and for the Polynesians. And stars and starlets have again become gods for Europeans as well as Americans.

There are also the histories of literature and of the visual arts, and, of course, of political and military power, of legal institutions and of economic change, to say nothing of their interrelationships.

All this, I suggest, points to a kind of historical pluralism: there is a plurality of cultural problems, of interests, and, perhaps most important, of individual characters and personal fates.

To conclude this section, I would like to make one additional remark. For you may well wonder how what I have said above relates to the criticisms that I have often given of the doctrine that the course of history is predictable, or that history has an intrinsic meaning.

Earlier I said that I *do not* believe that what I say here compromises those criticisms. But then, what *am* I doing, and what *is* my view?

I am here doing something that I would like to think is to be found in my work quite generally: when I have presented arguments against some view, I then see if there was not, perhaps, something of value which can be saved from the original position, whether a *corrective* should not perhaps be added to my criticism.[7] (This approach might well be described as 'dialectical'.)

Indeed, even in my *Poverty of Historicism*, in which my criticism of various historicist theses was first presented, I explicitly raised the question of whether there was not something, after all, in the historicist demand 'for a sociology which plays the role of a theoretical history, or a theory of historical development'.[8] And I there suggested that situational and institutional analysis (supplemented by the construction of models of political situations and social movements), on the one hand, and principles of historical interpretation, on the other, may serve to fill the gap created by the criticism of historicism.

Thus, what I have said above may be seen as an attempt, of a slightly different sort, to see whether we can – granted my criticisms of historicism – make anything of the historicist idea that there is an intrinsic plot in history.

I have suggested that one *can* say that the story of the growth of different kinds of human knowledge – and before it, of the evolution of animals and human life – is a plot that we can discover *in* history. But in saying this, I wish also to emphasize the improbability and fragility of these (progressive) developments. Not only was it highly improbable that things should have happened as they did, it also would have been all too easy for these developments to come to an end.

In this way I think that we can see, again, how the 'meaning' of history is something that we *choose*. For while this 'plot' – or, in view of the different kinds of knowledge, these 'plots' – are something given to us as a result of choices made by our forefathers, it is clearly up to us to make of them what we will. We can take them up, and foster them, or we can turn our backs on them. Certainly, no goddess of History will save us from the consequences of our own actions. And the fact that there may be weak biological tendencies in the direction of our plot is of little importance.

I need hardly add that if I suggest that we should foster this plot, I do so not on the grounds that it is good or desirable *because* it is there, but because this plot, and with it the motif of emancipation through knowledge, seem to me *worth* choosing and making our own.

II

I will now proceed to our second question. What is the use of history?

In an excellent paper entitled 'Philosophy of History before Historicism',[9] Professor George H. Nadel traces the history of the answers to this question. Foremost among these answers is what he calls the *Exemplar Theory of History*: the theory that history is of educational value, especially for the political education of the statesman or general.

'The Greeks are strong on precepts, the Romans are stronger on examples, which is a far greater thing', Nadel quotes from Quintilian. Polybius, incidentally, agrees, but turns it round: he alludes to Plato's demand that philosophers should become kings and kings philosophers, and he demands that not only should men of action become historians, but also historians men of action, for otherwise they will not know what they are writing about.

Under Stoic influence, history was regarded as a means of moral education – of education in righteousness.

This is a tradition that is still strong with Lord Acton, and its influence can be clearly felt in Sir Isaiah Berlin's famous lecture *Historical Inevitability* and, I suppose, also in my *Open Society*. Some of its strongest and wisest recent expressions may be found in the work of Ernst Badian on Roman and Hellenistic history.

Professor Nadel gives an outline of related theories. History, says Diodorus Siculus, restores the universal unity of mankind, a unity broken up by time and space. It thus ensures a kind of immortality and preserves the example of good men and good deeds.

Yet the exemplar theory declined. Hegel denied that statesmen actually learned from historical examples. Professor Nadel quotes from Hegel's *Philosophy of History*,[10] a passage that may be translated:

> It may be allowed that examples of virtue elevate the soul and are applicable in the moral instruction of children for impressing excellence upon their minds. But the destinies of people and states ... do not belong to this field. Rulers, statesmen, nations, are wont to be emphatically reminded of the teaching that experience offers in history. But what experience and history teach is this – that peoples and governments never learned anything from history, or acted on principles deduced from it.[11]

But the exemplar theory, while still powerful with Lord Acton, had actually been superseded before Acton by Acton's teacher, Leopold von Ranke (though Acton was by far closer to Döllinger than he was to Ranke). It was replaced, as Professor Nadel points out, by sheer professionalism: the view that history exists for its own sake – which really means that it exists for the sake of the historians. Nadel quotes the famous statement of Ranke's which is traditionally regarded as the manifesto of this position:[12]

> To history have been attributed the high offices of judging the past and instructing the present for the sake of the future. These high offices are beyond the aspirations of the present essay: it merely wants to show what actually happened.

This in brief is the story as Nadel tells it. But we need not give in to this approach of Ranke's any more than did Lord Acton. Again, I propose a pluralistic approach. History, I assert, can be interesting in itself. But it is interesting to the extent to which it tries to solve *interesting historical problems*. And some of these may be interesting because of our moral interests. Examples of such problems are: How did the two world wars break out? Were they avoidable?

The answers to these questions are certainly of great importance for the politician. *Pace* Hegel, a politician cannot be qualified for the Foreign Office unless he knows some of the historical facts and some of the historical conjectures concerning the Second World War. How guilty were 'The Appeasers'? What was the purpose of the Stalin purges? How was the decision to drop the two atom bombs on Japan arrived at?[13]

These are matters that ought to interest us all, even if we do not aspire to a post in the Foreign Office: these are intrinsically interesting historical problems, and of special interest if we want to *understand* the world in which we live.

But understanding the world we live in and ourselves is not everything. We also want to understand Plato, or Galileo, or Theodosius. And a good historian will add new fuel to this curiosity. He will make us wish to understand people, and situations, that we did not know about before.

III

With the word 'understand' I now come to the third and, I think, most interesting question: the question of the method of history, and especially the question of historical understanding.

For the last hundred years this question has been discussed very largely in terms of the difference in method between the natural sciences on the one hand, and the historical or humanistic sciences on the other. And there is an almost unanimous opinion that there is a big gulf between the two. There have been great quarrels about the details of this gulf. There are the famous German theorists, Windelband, Rickert, and Dilthey. There are the English theorists, foremost among them Collingwood. There is Professor Trevor-Roper, who objects to the intrinsic professionalism, and thus to the influence, of the natural scientist, and defends the view that history is for the layman. And there is Sir Isaiah Berlin, who warns us not 'to underestimate the differences between the methods of natural science and those of history or common sense'.[14]

I agree with Berlin's remark that the methods of history are those of 'common sense', and I always did agree with this view. I agree with Professor Trevor-Roper that nothing can be worse in history than a narrow professionalism, and I always did agree with this view. I agree with Collingwood and with Dilthey and with Hayek that we must try to *understand* historical events. And I agree that there is an urgent need for the philosopher of history to analyse, to explain, and indeed to understand historical understanding.

But my thesis has been for many years: *all those historians and philosophers of history who insist on the gulf between history and the natural sciences have a radically mistaken idea of the natural sciences.* They are not to be blamed for this: it is an idea fostered by the natural scientists themselves (and by positivist philosophers of science), and therefore, understandably enough, almost universally accepted. It has been greatly reinforced by the spectacular results of applied science. No wonder that so many philosophers and historians accept it.

It is, of course, undeniable that science has become the basis of technology. But what I regard as the true view of science is expressed on the jacket of a book by the great physicist and Nobel Laureate Sir George Thomson, one of the discoverers of the wave nature of the electron.[15] Thomson's book is called *The Inspiration of Science* – mark the title! – and the statement on the jacket begins with the words: 'Science is an art'. And it goes on to speak of the 'intrinsic beauty and wonder' of 'the ideas of modern physics'. Other great scientists have spoken in the same humanist vein, but few students of the humanities have taken them seriously. There are even some who go further, and believe, as I do, that the traditional profession-alist view of natural science is utterly mistaken. But with two exceptions, I have so far failed to convince any historian, or philosopher of history, that his intuitive idea of science is mistaken and that science is much more like history than historians think. The two exceptions are Professor Gombrich and Professor Hayek.

Professor Hayek, more especially, has for many years written against the emulation of the natural sciences by social scientists, including historians. He called the tendency to ape the methods of the natural sciences 'scientism'. Now I have been just as opposed to these scientistic tendencies as he. And I oppose them as much in the natural sciences as in the social sciences. For as I pointed out, more than twenty years ago, these 'scientistic' tendencies are in fact attempts to emulate what most people *mistakenly believe* to be the methods of the natural sciences, rather than the *actual* methods of the natural sciences. This view – that social scientists and philos-ophers of history have tried to imitate what they, quite mistakenly, believed to be the methods of natural science – has been most generously endorsed by Hayek in the preface to his book *Studies in Philosophy, Politics, and Economics*.[16]

But almost everyone else seems to be quite sure that the differences between the methodologies of history and of the natural sciences are vast. For, we are assured, it is well known that in the natural sciences we start from observation and proceed by induction to theory. And is it not obvious that in history we proceed very differently?

Yes, I agree that we proceed very differently. But we do so in the natural sciences as well.

In both we start from myths – from traditional prejudices, beset with error – and from these we proceed by criticism: by the critical elimination of errors. In both the role of evidence is, in the main, to correct our mistakes, our prejudices, our tentative theories – that is, to play a part in the critical discussion, in the elimination of error. By correcting our mistakes, we raise new problems. And in order to solve these problems, we invent conjectures, that is, tentative theories, which we submit to critical discussion, directed towards the elimination of error.

The whole process can be represented by a simplified schema which I may call the tetradic schema:

$$P_1 \rightarrow TT \rightarrow CD \rightarrow P_2$$

This schema is to be understood as follows. Assume that we start from some problem P_1 – it may be either a practical, or a theoretical, or a historical problem. We then proceed to formulate a tentative solution to the problem: a conjectural or hypothetical solution – a *tentative theory*, *TT*. This is then submitted to *critical discussions*, *CD*, in the light of evidence, if available. As a result, new problems, P_2, arise.

It should be said at once that this schema oversimplifies things. For there will be, in general, more than one problem to start with, and a multiplicity of conjectures will be offered as tentative solutions to every problem. There are also likely to be many different criticisms raised – especially if we test our conjectures by confronting them with observational evidence or with historical documentation. One could sum this up by saying that the schema should properly be fan-shaped: it should fan out towards the right.[17]

One other point is worth immediate comment. Since the schema is, as it were, self-propelling, starting with a problem and returning to a problem (although, of course P_1 and P_2 are not identical), it might well be said that we could start anywhere that we wanted to: that we could start just as well from tentative theories or from critical discussions as from problems. And the following argument could be put forward in favour of this view: problems arise, in general, against a background of knowledge – they presuppose a background of myths, of (tentative) theories or historical traditions. They also presuppose that these myths, theories and traditions are not accepted uncritically, but that some difficulties inherent in them have been noticed. Thus problems may be said to presuppose both tentative theories and critical discussion. On the other hand, Herodotus begins with a problem, and a modern historian like Lord Acton tells us to study problems instead of periods – that is, to start our study from a problem.

In fact, a case could be made out for each of the members of the triad *P*, *TT*, or *CD* as the starting point of science or of history. But although there is, from a logical point of view, little or nothing to choose between one or the other as a starting point, I prefer to say that we start from *problems*.

First of all, in saying that we begin with a problem and end with another problem, we point to one very important lesson: the lesson that the more our knowledge grows, the more we realize how little we know. This Socratic lesson is as true in the natural sciences as it is in history: to become educated is to get an inkling of the immensity of our ignorance.

At the same time, if we let our tetradic schema start with P_1, it allows us to say that it is the distance – often the great distance – between P_1 and P_2 that may serve as a measure of our progress in knowledge: the distance between the problem from which we started and the problem with which we are now faced.

A third reason in favour of the choice of P as our starting point is that we are often led to our researches by some practical problem that impressed itself upon us whether we wanted it or not. Thus modern economic theory might be said to have received much stimulus[18] from the currency crisis under William and Mary, from the distress at home, from William's urgent need for money (which reached its peak in 1696), and from the critical arguments – in support of the proposed stabilization of the currency, and against the counter-proposal of the Secretary of the Treasury to devalue the currency by 25 per cent – that were supplied by John Locke (and by Isaac Newton) and used by Montague in Parliament.[19] As so often, the problem from which the theory first sprang was a practical one. So were at least some of the famous problems of Archimedes. But as soon as a solution is offered, criticism takes over, and criticism is the engine of the growth of knowledge, as indicated by our tetradic schema.

It is of the utmost importance to realize that a bad problem and an erroneous conjecture are very much better than none. At the same time we must realize that this is so because we criticize our conjectures from the point of view of their adequacy – that is to say, their truth, their significance, and their relevance. That we constantly have their truth and their relevance in mind is perfectly compatible with the fact that many conjectures that may appear to us to be true at one stage may be discovered at a later stage to be erroneous. New documents may force us to reinterpret old documents. Or they may raise new problems. And in the light of a new problem an inscription that previously appeared insignificant may assume a completely unexpected significance.

This solves a famous, but, I think, not very deep methodological problem – the problem of historical relativism. Admittedly, our conjectures are relative to our problems, and our problems are relative to the state of our knowledge. And admittedly, there may be much in the momentary state of our knowledge that is erroneous. Yet this does not mean that truth is relative. It means only that the elimination of errors and the approach towards the truth are hard work. There is no criterion of truth. But there is something like a criterion of error: clashes arising within our knowledge or between our knowledge and the facts indicate that something is wrong. In this way knowledge can grow through the critical elimination of error. This is how we can get nearer to the truth.[20]

You will see that I can fully agree with Professor Trevor-Roper who, in his challenging and controversial inaugural lecture, pleads that we should keep the flow of ideas running from all tributaries, as he calls them, and especially from *lay tributaries*:[21]

> I myself believe that the historical contributions of both Sombart and Keynes are erroneous. I do not believe in the ... 'spirit of capitalism', nor do I believe that profit-inflation caused the expansion of sixteenth-century Europe and that we had Shakespeare when we could afford him. But what of that? These great tributaries which we have ignored have caused tremendous historical developments in other countries, and if we exclude them, we impoverish our own studies. They may be erroneous, but the mere correction of error involves first a new study, and then a new interest which that error has created. In humane studies there are times when a new error is *more* life-giving than an old truth, a fertile error than a sterile accuracy.

I agree with Professor Trevor-Roper, except on one point: his apparent belief[22] that what he says holds only for the humane studies, and not for the natural sciences. Admittedly, specialists are needed in both the sciences and the humane studies. But specialization and the professionalist attitude of superiority and exclusiveness towards the outsider or layman must lead to the drying up of both humanist and scientific studies.

Professor Elton in his book, *The Practice of History*, defends professionalism. But does it need defence? Did Ranke not win this battle a hundred years ago? It seems to me, rather, that it has become necessary now to remind the great professionalists, and specialists, whether in history, in science, or in medicine, that they too are liable to make mistakes – that is, professional mistakes. It is regarded as wrong and even humiliating to make mistakes. But who has not made them? The historian may think that a great physicist makes no mistakes in his subject. But were he to study the history of physics he would soon find that even the greatest physicists have made mistakes. Einstein worked from 1905 to 1915 on the problem of

gravitation before he arrived at a theory that could replace Newton's, spending almost the whole of the last three of these ten years
on what he described as a completely mistaken track. And even after
he had found his field equations, he was told by Kretschmann in
1917 that what he had proposed as an essential argument was
mistaken. Einstein at once admitted the mistake. But what he then
said in order to replace his argument (he hinted that Newton's
equations could only with considerable difficulty be put in a
covariant form) was again mistaken, as has been shown since.[23]

Nobody is exempt from making mistakes. The great thing is to
learn from them. And this is done by criticism, and by the discovery
of new *problems* brought forth by criticism.

I think that this is implicitly admitted in Elton's book. He
distinguishes between historical analysis – the analysis of historical
problems – and historical narrative. Yet he argues against Lord
Acton's excellent advice to young historians, given in his inaugural
lecture of 1895,[24] that they should '*study problems in preference to
periods*'.

Lord Acton's views on method, like Collingwood's or Professor
Trevor-Roper's, can, I think, be shown to be essentially in agreement
with those I am defending here. Elton, however, seems to dislike
them. But a close reading of what he says shows that in the end he
seems to agree with Lord Acton. Let me quote Elton:

> To study problems not periods was Lord Acton's much-
> quoted injunction, and those who cite him approvingly fail to
> note that it is now some seventy years since he uttered those
> gnomic words, and that in actual fact he proved incapable of
> studying either problems or periods to a practical conclusion.
> The historian, working in the records and meeting one
> unresolved problem after another, quite naturally persuades
> himself that the real work consists in tackling these dark
> entities.

– that is, the problems. 'But that does not mean that one should pin
a special medal on analysis',[25] Elton adds – which means, apparently,
that one should not be specially concerned with problem solving. So
far, it will be seen, no argument has been brought forward against
Acton except that his words are 'gnomic' and seventy-two years
old.[26] Yet Elton's next two sentences are in fact an admission that
Acton is right. The first reads: 'Since history is the record of events,
and of problems, proceeding through time, narrative must be not

only legitimate but also urgently called for.' In this sentence 'problems proceeding through time'[27] are appealed to. This will hardly do as an argument against Acton's stress on problems, for Acton never said that you should not follow your problems through time. The next sentence of Elton's is even clearer: 'Once again, the only point which determines the choice is the historian's purpose, the questions he is asking.'[28] I entirely agree. *The questions* the historian is asking are decisive. But 'the questions the historian is asking' is merely a synonym for the term 'historical problem'. And so we are back at Lord Acton's emphasis on problems.

In fact, it seems that our work can start only from problems. And this holds true not only for what Elton calls 'analysis' but just as much for what he calls 'narrative'.

It may perhaps be useful to point out here that Leopold von Ranke's famous professionalist revolution of history had in it more than a streak of what Hayek has called 'scientism'. The alleged method of the professional scientist is: start from observations, observe, and go on observing. The alleged method of the professional historian is: start from documents, read documents, and go on reading documents.

These alleged methods are exactly analogous, and both are precepts which cannot be carried out: they are logically impossible. You cannot start from observation: you have to know first *what to observe*. That is, you have to start from a problem.[29] Moreover, there is no such thing as an uninterpreted observation. All observations are interpreted in the light of theories. Exactly the same holds for documents. Is my train ticket to London a historical document? Yes and no. If I am accused of murder, the ticket may possibly serve to support an alibi, and so become an important historical document (as in Dorothy Sayers' *Five Red Herrings*). Nevertheless, I should not advise a historian to start his work by collecting used railway tickets.

A historical document, like a scientific observation, is a document only relative to a historical problem. And like an observation, it has to be *interpreted*. This is one of the reasons why people may be blind to the significance of a document, and destroy it. Or why they may destroy (as Elton complains) the order of some documents and, with it, one of the clues to their interpretation.

So far I have tried to produce a few arguments to show that there is more in common between the actual method of science and the actual method of history than most historians realize. The similarity extends even to the scientistic misinterpretations of the two methods, as my last remarks show.

But is there not a fundamental difference: one connected with *the problem of understanding history*?

I will outline very briefly Collingwood's theory of sympathetic understanding or, as it may be called, empathy, which we find in his posthumously published book *The Idea of History*. Collingwood's theory may be stated in brief as follows: historical knowledge, or historical understanding, consists in the *re-enactment by the historian of past experience*. Let me quote a passage from Collingwood, a passage with which I agree to a considerable extent.

> Suppose [a historian] is reading the Theodosian Code, and has before him a certain edict of an emperor. Merely reading the words and being able to translate them does not amount to knowing their historical significance. In order to do that he must envisage the situation with which the emperor was trying to deal, and he must envisage it as that emperor envisaged it. Then he must see for himself, just as if the emperor's situation were his own, how such a situation might be dealt with; he must see the possible alternatives, and the reasons for choosing one rather than another; and thus he must go through the process which the emperor went through in deciding on this particular course. Thus he is re-enacting in his own mind the experience of the emperor; and only in so far as he does this has he any historical knowledge, as distinct from a merely philological knowledge, of the meaning of the edict.
>
> Or again, suppose he is reading a passage of an ancient philosopher. Once more, he must know the language in a philological sense and be able to construe; but by doing that he has not yet understood the passage as an historian of philosophy must understand it. In order to do that, he must see what the philosophical problem was, of which his author is here stating his solution. He must think that problem out for himself, see what possible solutions of it might be offered, and see why this particular philosopher chose that solution instead of another. This means re-thinking for himself the thought of his author and nothing short of that will make him the historian of that author's philosophy.[30]

What Collingwood here describes I have tried to describe in *The*

Poverty of Historicism, and in *The Open Society*, and in other works, under the name of *situational logic* or *situational analysis*.[31] What we have to do, I suggested, was to reconstruct the *problem situation* in which the acting person finds himself, and to show how and why his action constituted a solution of the problem as he saw it.

Yet I have mentioned above that I agree with Collingwood's passage only to a considerable extent. Why not entirely?

There *is* a difference between Collingwood's theory and mine. It seems a small difference, but it has far-reaching consequences.

The difference is this. Collingwood makes it clear that the essential thing in understanding history is not so much the analysis of the situation as the historian's mental process of re-enactment. The analysis of the situation serves only as an indispensable aid for this re-enactment.[32] I, on the other hand, suggest that the psychological process of re-enactment is inessential, though I admit that it can greatly help the historian, by providing a kind of intuitive check of the success of his situational analysis. What is essential, I suggest, is not the re-enactment but the situational analysis: the historian's attempt to analyse and describe the situation is nothing else than his historical conjecture, his historical theory. And the question – 'What were the important or operative elements in the situation?' – is the central problem that the historian tries to solve. To the extent that he solves it, he has *understood* the historical situation and the piece of history that he tries to recapture.

What he has to do *qua* historian is not to re-enact what has happened, but to give objective arguments in support of his situational analysis. He may well be able to do so, while the re-enactment may or may not work. For the act may be in many ways beyond him. It may be an act of cruelty or an act of heroism, which he may be unable to re-enact. Or it may be an achievement in art, or literature, or science, or philosophy, that exceeds his abilities. Yet all this does not prevent him from making interesting historical discoveries – from finding new solutions to old historical problems, or even from finding new historical problems.

The main significance of the difference between Collingwood's re-enactment method and my method of situational analysis is that Collingwood's is a *subjectivist* method, while the method I advocate is *objectivist*.[33] It would seem that if we follow Collingwood, *a systematic rational criticism of competing solutions to historical problems becomes impossible.* For we can rationally criticize only conjectures or theories that have not become part of ourselves, but can be put outside ourselves, and may thus be inspected by everybody, especially by those who hold different theories. The objectivist method of situational analysis, on the other hand, permits the *critical discussion of our tentative solutions* – of our attempts to reconstruct the situation. And to this extent it is, indeed, much nearer to the actual method of the natural sciences.

Let me take a very simple example. It is well known that Galileo was unwilling to accept the lunar theory of the tides, and that he made tremendous efforts trying to explain the tides by a non-lunar theory. It is also known that Galileo did not reciprocate the friendly advances of Kepler. These two facts create two problems. And they may give rise to the following explanatory historical conjecture: Galileo was opposed to astrology – to the theory that the positions of the planets, including the moon, have an influence upon terrestrial events. Documents show that the lunar theory of the tides was indeed part of astrological lore. And the fact that Kepler was a professional astrologer was, of course, known to Galileo.

Now a rereading of Galileo's *Dialogue Concerning the Two Chief World Systems* with this conjecture in mind led me to the following passage (the last one in which Kepler is mentioned), which seems to corroborate the conjecture:[34]

> Thus everything that has been previously conjectured by others [concerning the explanation of the tides] seems to me completely invalid. But among all the great men who have philosophized about this remarkable effect, I am more aston- ished at Kepler than at any other. Despite his open and acute mind, and though he has at his fingertips the motions attributed to the earth, he nevertheless lent his ear and his assent to the moon's domination over the waters, to occult properties, and to such puerilities.

It so happens that when I had previously read this passage, I did not understand the full weight of the reference to 'occult properties': it was only after I had been troubled by these two problems, and had produced my conjecture, that I understood this passage fully.

It is clear that this little bit of very simple historical problem solving operates with what I call situational logic or situational analysis. This method of analysis helps us to explain two of Galileo's attitudes – one towards a scientific problem, one towards a person – by way of a conjectural reconstruction of the problem situation, as he may have seen it. Yet this reconstruction is not a real re-enactment in Collingwood's sense. It is not a re-enactment of Galileo's thoughts and actions that is of interest here. Neither is it a reinvention of Galileo's theory of the tides (an act of which I am quite incapable). Nor is it a re-enactment of his failure to reply to some of Kepler's letters (although failing to reply to a letter or even to two letters is a thing of which I am perfectly capable).

Now Galileo's failure to reply to Kepler is clearly one of those things simply not worth re-enacting: in itself it is too trivial an action (or, rather, inaction). But as a symptom, and in connection with another historical problem, it may be interesting. And it is so from the point of view of our situational diagnosis.

I therefore claim for situational analysis that it is a better theory of historical understanding than Collingwood's re-enactment theory. It is less rigid. It is not confined, as is Collingwood's theory, to the re-enactment of conscious thought processes, but makes allowance for the reconstruction of problem situations that were incompletely understood by the agent. Furthermore, it makes room for the reconstruction and analysis of situations that arise as the unintended and unforeseen consequences of our actions – a very important point indeed. And it allows us to give full weight, in our situational analysis, not only to individuals but also to institutions. In other words, it is wider, or, as I may say, far more pluralistic than even that of Collingwood who, by his strong emphasis on problems, approached history in a far more pluralistic spirit than any of his predecessors. For Collingwood, the re-enactment of any thought may become a problem. For situational logic, the reconstruction of any situation, including the reconstruction of the one that brought about another, may become a problem. Moreover, situational logic is as much concerned with the situation as experienced by the acting subject as with the objective situation as it actually was, and thus with the objective errors of the acting subject.

This brings me to the most important difference between my approach and Collingwood's. For Collingwood, as for almost all philosophers, knowledge consists, essentially, in living experiences of the knowing subject. And this, of course, holds for historical knowledge. For me, knowledge consists essentially of *exosomatic artefacts*, or products, or institutions.[35] (It is their exosomatic character that makes them rationally criticizable.) There is knowledge without a knowing subject – that knowledge, for example, which is stored in our libraries. Thus there can be growth of knowledge without any growth of awareness in the knower. The growth of knowledge can even form the main plot of our history. And yet there *may* be no corresponding increase in either our subjective knowledge or our abilities. There may even be no change in our interests. Human knowledge may grow outside of human beings.

Thus it is possible to differentiate between the evolution of man (in the singular), that is, of mankind and its exosomatic knowledge, and the history of different individual men (in the plural). And there is no doubt in my mind that the main value and the main characteristic of the subject that we call history, indeed of all humanistic subjects, is that it is wide enough to interest itself not only in the evolution of the human race and its institutions, but in the stories of individual men (in the plural) and their struggles with their institutions, with their evolving environment, and with the problems posed by the evolution of man and his knowledge.

Thus history is pluralistic. It deals not only with man but with men. Above all it allows us to raise the problem how much or how little the growth of knowledge, the history of art, and the evolution of man has affected men. This problem, I suggest, is one of the greatest problems of history.

NOTES

1 Cp. the opening section of my 'Science: Problems, Aims, Responsibilities', chapter 4 of the present volume.

2 The last paragraph closely follows p. 346 of my *Conjectures and Refutations*.

3 Cp. my *Poverty of Historicism*, pp. 3f. and 17, my *Open Society*, volume II, pp. 208 and 255–6, and the interesting discussion in Alan Donagan's paper 'Popper's Examination of Historicism', in *The Philosophy of Karl Popper*, edited by P.A. Schilpp, in The Library of Living Philosophers, The Open Court Publishing Co., La Salle, Illinois, 1974, pp. 905–24.

4 For a discussion of historicist theories in the arts, and the undesirable consequences to which they have led, see my 'Intellectual Autobiography', especially sections 13 and 14, and Ernst Gombrich, 'The Logic of Vanity Fair', both in *The Philosophy of Karl Popper*, edited by P.A. Schilpp. My 'Intellectual Autobiography' is now independently published under the title *Unended Quest*.

5 See Gombrich's paper cited in note 4, and my reply on pp. 1174–80 of the same book.

6 Cp. my discussion in chapter 2 above.

7 An example of this may be found in the development of my ideas regarding the so-called 'empirical basis' of knowledge. Contrary to the commonsensical idea that our perceptions are *given* to us by the world, I have stressed, as a corrective, the role of our active participation:

'Making comes before matching'. But this is itself in need of a further corrective. For if taken systematically, it would lead to idealism, with reality as our construction. So I then introduce the corrective idea that we get into contact with reality through an empirical refutation, much as does a man when he hits a brick wall.

8 See my *Poverty of Historicism*, especially the first paragraph of section 31.

9 George H. Nadel, 'Philosophy of History before Historicism', *The Critical Approach to Science and Philosophy*, edited by Mario Bunge, The Free Press, Glencoe, Illinois, 1964, pp. 445–70.

10 Cp. Nadel, p. 469, note 2.

11 G.W.F. Hegel, *The Philosophy of History*; cp. Sibree's translation, 1956, p. 6.

12 Leopold von Ranke, *Geschichte der Romanischen und Germanischen Völker*, (1824), 3rd edition 1885, p. vii (my translation).

13 See also chapter 6 of the present volume.

14 Isaiah Berlin, *Historical Inevitability*, Auguste Comte Memorial Trust Lecture, Oxford University Press, London, 1954; see p. 11, footnote.

15 George Thomson, *The Inspiration of Science*, Oxford University Press, London, 1961.

16 F.A. von Hayek, *Studies in Philosophy, Politics, and Economics*, Routledge & Kegan Paul, London, 1967, p. viii. (The reader may be reminded that the present essay was first published as a contribution to a Hayek *Festschrift*, and that the lecture on which it was based was given in 1967.)

17 Cp. my *Objective Knowledge*, pp. 243 and 287.

18 The following comment in the Preface of W. Letwin's *The Origins of Scientific Economics*, Methuen & Co., 1963, p. ix, is most interesting in this connection: 'They [the inventors of economic theories at the end of the seventeenth century] created scientific theories, yet they generally did not do so deliberately, nor did they do it even for the sake of knowledge, but rather their scientific accomplishments were a by-product of their efforts to convince others to accede to certain economic policies. To show how these practical, and often mercenary, objectives led certain men to build a new science, the first social science, is my chief purpose.'

19 This example is based on the account in Macaulay's *History of England*, chapter XXI. For a more recent discussion of the currency crisis and the various economic theories that were put forward in the course of the controversy, see W. Letwin, *The Origins of Scientific Economics*, pp. 64–75 and 166–71, and J.K. Horsefield, *British Monetary Experiments 1650–1710*, G. Bell & Sons, London, 1960, pp. 23–70. (According to Horsefield, Newton's memorandum on the currency crisis was in favour of devaluation.)

20 Only the relativity of *truth* is contested here. *Relevance* is relative, but this does not give rise to a problem of historical relativity. There can be no real clash between different claims to relevance: this is just one of the reasons for a pluralist approach to the philosophy of history.

21 Hugh R. Trevor-Roper, *History: Professional and Lay*, Clarendon Press,

Oxford, 1957, pp. 21f.

22 See, for example, p. 13: '... the view which I wish to express springs from the conviction that history is a humane study and that the study of the humanities requires a different method from the study of the sciences.'

23 Peter Havas, 'Four-Dimensional Formulations of Newtonian Mechanics and Their Relation to the Special and the General Theory of Relativity', *Review of Modern Physics*, 36, 1964, pp. 938–65.

24 Lord Acton, *Inaugural Lecture on the Study of History*, London, 1895. Cp. his *Lectures on Modern History*, 1906, or *Essays in the Liberal Interpretation of History*, edited by W.H. McNeill, University of Chicago Press, Chicago and London, 1967, pp. 350f.

25 Cp. G.R. Elton, *The Practice of History*, Sydney University Press, Sydney, 1967, p. 127.

26 It can hardly be said to be a criticism that Acton did not succeed in carrying his somewhat optimistic plans to a 'practical conclusion'.

27 Elton, pp. 127f.

28 Elton, p. 128.

29 This was seen clearly by Ranke's contemporary, Gustav Droysen. In his lectures on historical method he says: 'For research does not proceed by random discoveries; it *seeks* something. It must know what it is looking for if it is to find anything at all.' Johann Gustav Droysen, *Historik: Vorlesungen über Enzyklopädie und Methodologie der Geschichte*, edited by Rudolf Hübner, 1936, p. 35. (I owe this reference to Kims Collins.)

30 R.G. Collingwood, *The Idea of History*, Oxford University Press, London, 1946, p. 283.

31 See my *Poverty of Historicism*, pp. 149f.; my *Open Society*, volume II, pp. 97 and 265; chapter 4 of my *Objective Knowledge*, and chapter 8 of the present volume.

32 Since this paper was first published in *Roads to Freedom*, Margit Hurup Nielsen has drawn my attention to the fact that my interpretation of Collingwood is perhaps controversial, in that it seems now to be widely accepted that Collingwood, when talking of re-enactment, was talking of a reconstruction, by the historian, of something close to what I would call the objective thought contents of the historical agent, as distinct from a re-experiencing of his feelings. If this interpretation is correct, then Collingwood's views are closer to mine than I had supposed. But there are nevertheless important points at issue between us, some of which are brought out in my text following note 34.

33 More precisely, my theory is an objectivist theory of subjective understanding. Thus I share with the 'subjectivist' approach an emphasis on the situation as understood by the agent, and reject attempts to explain human action in 'objective' terms where 'objective' means 'behaviourist' or 'physicalist'. For this objectivist theory of understanding, see chapter 4 of my *Objective Knowledge*.

34 Galileo Galilei, *Dialogue Concerning the Two Chief World Systems*, translated by Stillman Drake, University of California Press, Berkeley and Los Angeles, 1967, p. 462. (Cp. my *Objective Knowledge*, p. 173.)

35 Since this lecture was first published I have discussed this idea in much

more detail. See particularly chapters 3, 4, and 8 of my *Objective Knowledge*, sections 38f. of my *Unended Quest*, sections 20–6 of my replies in *The Philosophy of Karl Popper*, and chapter P2 of *The Self and Its Brain*. What I have here called exosomatic artefacts I now usually call World 3.

8

MODELS, INSTRUMENTS, AND TRUTH

The status of the rationality principle in the social sciences

When I received the invitation to present to you my views on the methodology of the social sciences, I felt very honoured indeed. But I also felt a little uneasy, for the following reasons. My views on the methodology of the social sciences are the result of my admiration for economic theory: I began to develop them, some twenty-five years ago, by trying to generalize the method of theoretical economics.[1] You will understand my fear that you may, as economists, find my views trivial – if not altogether out of date.

It was this fear which made me decide to devote about one-third of this lecture to my views on the methodology of science in general, one third (sections 2 to 7) to problems peculiar to the methods of social science, and the remainder (sections 8 to 11) to an attack upon the instrumentalist philosophy of science – that still fashionable philosophical theory of pragmatism which tells us that our theories are nothing but instruments. To this I shall oppose my own view according to which theories are steps in our search for truth – or to be both more explicit and more modest, in our search for better and

Based on a lecture delivered in the Department of Economics, Harvard University, on 26 February 1963. The lecture was revised after it was delivered and two new sections (12 and 13) were added in 1963 and 1964. An extract of this paper has been published in French, under the title '*La rationalité et le statut du principe de rationalité*' in *Les fondements philosophiques des systèmes économiques: Textes de Jacques Rueff et essais rédigés en son honneur*, edited by E.M. Claassen, Payot, Paris, 1967, pp. 142–50. The paper has been left unchanged, apart from certain minor corrections, the addition of the notes, and a few other additions as indicated in the notes.

better solutions of deeper and deeper problems (where 'better and better' means, as we shall see, 'nearer and nearer to the truth').[2]

1. Problems, theories, and criticism

The views on the methods of the social sciences that I am going to sketch are briefly these. The methods appropriate to the social sciences are totally different from the methods of the natural sciences *as they are usually described by textbooks, by tradition, and by the majority of natural and social scientists*. But this is so merely because all these textbooks and these traditions and these scientists are totally mistaken about the methods of the natural sciences. Once we get a proper understanding of the methods of the natural sciences, we can see that there is a great deal in common between them and the methods of the social sciences.

The main misunderstanding about the natural sciences lies in the belief that science – or the scientist – starts from observation and the collection of data or facts or measurements, and thence proceeds to connect or correlate these, and so to arrive – somehow – at generalizations and theories.

I remember an occasion when I was chairman of a meeting in which a distinguished scientist presented this view. Science, he said, was just measuring, and correlating results. In the discussion that followed, I suggested that we should ask for a grant for a project of measuring the length, width, thickness, and weight of the books in the British Museum – in order to study possible correlations between these measurements. I predicted that we should be able to find strong positive correlations between the product of the first three measures and the fourth.

Why is this project absurd? Because it is uninteresting. Because it starts from the collection of data rather than from a scientific *problem*. And because there is no reason to think that it will throw light on the more pressing scientific problems of the day.

The work of the scientist does not start with the collection of data, but with the sensitive selection of a promising *problem* – a problem that is significant within the current *problem situation*, which in its turn is entirely dominated by our theories.

It is my view that the methods of the natural as well as the social sciences can be best understood if we admit that *science always begins and ends with problems*. The progress of science lies, essentially, in the evolution of its problems. And it can be gauged by

155

the increasing refinement, wealth, fertility, and depth of its problems.

Scientific problems are preceded, of course, by pre-scientific problems, and especially by practical problems. Even the amoeba, we may safely assume, has problems. For every organism has built-in expectations. And problems arise, most characteristically, when some of these expectations are disappointed.

You may ask how it is possible for us to begin with problems, and how there can possibly be any problems in the absence of all prior knowledge – for example, in the form of expectations. This question is very much to the point. And my answer is that we never start afresh, from nothing, so to speak, with a completely innocent mind. The growth of knowledge always consists in correcting earlier knowledge. Historically, science begins with pre-scientific knowledge, with pre-scientific myths and pre-scientific expectations. And these, in turn, have no 'beginnings'. They 'begin' when life begins. And even at the beginning of life there are problems – problems of survival. Thus no first knowledge was ever engraved on an innocent mind, or on a *tabula rasa*, or an empty slate. There is simply no new knowledge without some kind of earlier knowledge, some kind of expectation, upon which it is a modification. And such modifications occur especially when earlier knowledge runs into trouble – for example, when an expectation is disappointed, when it gives rise to a *problem*.

Thus we can look upon any particular item of knowledge, and especially upon any scientific theory, as a tentative solution to some problem or other, and as giving rise to new problems. And the fertility and depth of our theories may well be measured by the fertility and depth of the new problems to which they give rise.

As I have admitted that every problem arises out of some kind of knowledge, and therefore presupposes knowledge, you may ask whether my remark that science begins and ends with problems could not be replaced by the remark that science begins and ends with knowledge. My answer is, 'Yes – provided you mean (as I do) by *knowledge* something like problematic or hypothetical or tentative knowledge rather than unproblematic or settled knowledge.' Settled knowledge does not *grow*. I have often said myself that science begins and ends with theories. But I used the term 'theory' in a very wide sense, in a sense in which it includes myths and all kinds of expectations and guesses. I never use it in the sense of a settled or proved or established theory, for I do not think that such a theory exists. A theory always remains hypothetical, or conjectural. It always remains guesswork. And there is no theory which is not beset with problems.

Yet I think that the remark that science begins and ends with problems is perhaps a little more informative than the remark that science begins and ends with theories.

To see this let us consider for a moment what it means to *understand* a theory.

Understanding a theory means, I suggest, understanding it *as an attempt to solve a certain problem*. This is an important proposition, and one which too few people understand. The problem which a theory is intended to solve may be either a practical problem (such as finding a cure for, or a preventative against, poliomyelitis or inflation) or a theoretical problem – that is, a problem of explanation (such as explaining how poliomyelitis is transmitted, or how inflation comes about).

What is the point of, say, Newton's theory? It is *an attempt to solve the problem of deriving, and, to that extent, of explaining* Kepler's laws *and* Galileo's laws. (I shall not go here into the question of why Newton himself did not regard his theory as explanatory.) Without understanding the problem situation which gave rise to the theory, the theory is pointless – that is, it cannot be properly understood. Similarly, without understanding the problems raised by depression and unemployment for neo-classical economics, Keynes' theory must appear pointless, and cannot be fully understood. It can only be understood *as an attempt to solve these problems*. You see from this that at least from the point of view of *understanding science* – that is, understanding its theories – problems are prior to theories. This is one of the reasons why I believe that in saying that *science begins and ends with problems* I am giving you a simple formula of considerable power and applicability.

Of course, we must now ask: What is it to understand a problem? If a young scientist is supposed to *begin* with a problem, how can he ever be in the position to understand it? How, therefore, can he ever begin with a problem?

My answer is that there really is only one way of *learning to understand a problem* which we do not yet understand – and that is to try to solve it *and to fail*.

This may sound paradoxical. For how can we try to solve a problem which we do not even understand?

157

The answer to this question is that if we do not understand the problem, we shall certainly – or almost certainly – not solve it. But the certainty of failure need not prevent us from trying.

Take as an example a practical problem, such as learning to ride a bicycle or to play the violin. With the exception perhaps of a few geniuses, all those who do not yet understand the problem of riding a bicycle are likely to fail at their first attempts to solve it. And so are those who do not yet understand the problem of playing the violin. But after a few failures they may begin to appreciate where the difficulty lies: they will begin to understand the problem. For a problem is nothing but a difficulty. And to understand a problem is nothing but being aware of what this particular difficulty is like.

Practical problems may give rise to theoretical problems. For example, the practical problem of riding a bicycle may give rise to the theoretical problem of explaining how and why the rider keeps his balance. And the practical problem of playing a musical instrument, or of making one, may give rise to the development of the theory of acoustics. Now in every such theory, new problems will arise again and again. These may be perhaps internal difficulties of the theory, such as explanations which we find for some reason unsatisfactory, or clashes between the theory and the facts. The theory evolves as a result of our attempts to solve these problems.

One can say that most theoretical problems arise from cases in which the theory has let us down, so that some kind of repair of the theory is needed. But this is precisely the reason why we do not, in general, 'understand' the new problem at first: our theory (that is, the old theory, the one of which we know something) is insufficient, and we do not know what is wrong with it. But we can learn to understand the problem better and better by trying to adjust or repair our theory, or to replace it by another one. Clearly, these attempts are not likely to succeed as long as we do not 'understand' the problem before us. But my thesis is that by criticizing our own attempts – our failures – we learn more and more about our problem: we learn where its difficulties lie. As with practical and pre-scientific problems, we learn from our mistakes, from our failures, by something like a feedback method.

My view of the method of science is, very simply, that it systematizes the pre-scientific method of learning from our mistakes. It does so by the device called *critical discussion*.

My whole view of scientific method may be summed up by saying that it consists of these four steps:

1   We select some *problem* – perhaps by stumbling over it.
2   We try to *solve* it by proposing a *theory* as a tentative solution.
3   Through the *critical discussion of our theories* our knowledge grows by the elimination of some of our errors, and in this way we learn to understand our problems, and our theories, and the need for new solutions.
4   The critical discussion of even our best theories always reveals new problems.

Or to put these four steps into four words: *problems – theories – criticisms – new problems*.

Of these four all-important categories the one which is most characteristic of *science* is that of error-elimination through *criticism*. For what we vaguely call the *objectivity of science*, and the *rationality of science*, are merely aspects of the *critical discussion* of scientific theories.

To see this, it is important to be clear about the aims of the critical discussion of a scientific theory. Criticism of a scientific theory is always an attempt to find (and to eliminate) a mistake, a flaw, or an error *within* the theory. It is, as I have indicated, the negative feedback by which we control the construction of our theories. It tries to show either that the theory has unacceptable consequences, or that it does not solve the problem it sets out to solve, or that it merely shifts the problem, raising difficulties worse than it surmounts, or that it is inferior to some of the competing theories – that it is, for example, weaker, or more complex.

This is the aim of scientific criticism. It is important to note what scientific criticism does *not* try to show. It does not try to show that the theory in question has not been proved or demonstrated. Similarly, it does not try to show that the theory in question has not been established or justified – *because no theory can be established or justified*. Incidentally, it does not try to show that the theory in question has a high probability (in the sense of the probability calculus) – *because no theory has a high probability* (in the sense of the probability calculus).

Accordingly, scientists, in their critical discussions, do not attack the arguments which might be used to establish, or even to support, the theory under examination. They attack the theory itself, *qua* solution of the problem it tries to solve. They examine and challenge its consequences, its explanatory power, its consistency, and its compatibility with other theories.

What we call *scientific objectivity* is nothing else than the fact that no scientific theory is accepted as a dogma, and that all theories are tentative and are open all the time to severe criticism – to a rational critical discussion aiming at the elimination of errors.

As to the *rationality of science*, this is simply the rationality of critical discussion. Indeed there is nothing, I think, which can better explain the somewhat abstract idea of *rationality* than the example of a well-conducted critical discussion. And a critical discussion is well-conducted if it is entirely devoted to one aim: to find a flaw in the claim that a certain theory presents a solution to a certain problem. The scientists participating in the critical discussion constantly try to refute the theory, or at least its claim that it can solve its problem.

It is most important to see that a critical discussion always deals with more than one theory at a time. For in trying to assess the merits or demerits even of *one* theory, it always must try to judge whether the theory in question is an *advance*: whether it explains things which we have been unable to explain so far – that is to say, with the help of older theories. But of course there is often (in fact, always) more than one new theory competing at a time, in which case the critical discussion tries to assess their comparative merits and demerits. Older theories, however, always play an important part in the critical discussion, especially those theories which form part of the 'background knowledge' of the discussion – theories which, for the time being, are not criticized, but are used as the framework within which the discussion takes place. Any single one of these background theories may however be challenged at any time (though not too many at the same time), and thus move into the foreground of the discussion. Though there is always a background, any particular part of the background may at any time lose its background character.

Thus critical discussion is essentially a comparison of the merits and demerits of two or more theories (usually more than two). The merits discussed are, mainly, the *explanatory power* of the theories (discussed at some length in my *Logic of Scientific Discovery*) – the way in which they are able to solve our problems of explaining things, the way in which the theories cohere with certain other highly valued theories, their power to shed new light on old problems and to suggest new problems. The chief demerit is inconsistency, including inconsistency with the results of experiments that a competing theory can explain.

It will be seen from this that critical discussion will often be undecided, and that there do not exist very definite criteria for tentative acceptability: that the frontier of science is very fluid.

Thus the result of a scientific discussion is very often inconclusive, not only in the sense that we cannot conclusively verify (or even falsify) any of the theories under discussion – this should by now be obvious – but also in the sense that we cannot say that one of our theories seems to have definite advantages over its competitors. If we are lucky, however, we may sometimes come to the conclusion that one of the theories has greater merits and lesser demerits than the others. (In this case some people say that the theory is 'accepted' – of course, only for the time being.)

From this analysis of the process of the critical discussion of theories it should be clear that the discussion never considers the question whether a theory is 'justified' in the sense that we are justified in accepting it as *true*. At best, the critical discussion justifies the claim that the theory in question is the best available, or, in other words, that it comes nearest to the truth.

Thus although we can judge theories only 'relatively' in the sense that we compare them with each other (and not with the truth, which we do not know), this does not mean that we are relativists (in the sense of the famous phrase that 'truth is relative'). On the contrary, in comparing them, we try to find the one which we judge comes nearest to the (unknown) truth. So the idea of truth (of an 'absolute' truth) plays a most important part in our discussion. It is our main regulative idea. Though we can never justify the claim to have reached truth, we can often give some very good reasons, or justification, why one theory should be judged to be nearer to it than another.

What I have said so far is meant to apply equally to the natural and to the social sciences. At this stage I will add only one remark which could have some bearing on the question of the differences – or the alleged differences – between them.

One of the most telling and important forms of criticism – that is, of the critical discussion of theories – is the appeal to observation, experiment, and measurement. If we can show that the consequences of a theory do not agree with certain facts (or with certain observations or measurements), then we have a strong argument against it. We may even kill it, especially if we can show that the

161

falsifying experiment can be explained by some competing theory. But observations, experiments, and measurements are of interest only in the context of the critical discussion of some theory. They are neither starting points of science, nor data.

Nevertheless, observations, experiments, and measurements can, by refuting some accepted theory, create a new problem, and thus start a new line of development. And a falsifying experiment is one of the characteristic ways in which new problems arise in the empirical sciences. But there are other characteristic ways. Internal difficulties, for example, may be detected within a theory. Or we may have dealt with various problems quite successfully, solving each of them by a different theory, only to find that some of these theories are not mutually compatible. While some people may accept this situation, others may see an important problem here – the problem of finding a reconciliation or, preferably, of finding a new and more comprehensive theory.

But before proceeding to discuss the question of the peculiarities of the social sciences, I wish to repeat that the only [3] function which my theory of method assigns to observations, experiments, and measurements is the modest although important one of assisting criticism – that is, assisting in the discovery of our mistakes.

With this, I conclude my very general remarks on what I believe to be the critical methods common to both the natural and the social sciences, and I now proceed to some points which will help to bring out some of the peculiarities of the methods of the social sciences.

2. Models and situations

In this second part of my lecture, I shall try to explain some of the similarities between the natural and the social sciences and also some of the dissimilarities.

Let me begin by distinguishing between two kinds of problems of explanation or prediction.

1 The first kind is that of explaining or predicting one or a smallish number of *singular events*. An example from the natural sciences would be, 'When will the next lunar eclipse (or, say, the next two or three lunar eclipses) occur?' (An example from the social sciences would be, 'When will there be the next rise in the rate of unemployment in the Midlands, or in Western Ontario?')

2   The second kind is that of explaining or predicting a certain *kind or type* of event. An example from the natural sciences would be, 'Why do lunar eclipses occur again and again, and only when there is a full moon?' (An example from the social sciences would be, 'Why is there a seasonal increase and decrease of unemployment in the building industry?')

The difference between these two kinds of problems is that the first can be solved *without constructing a model*, while the second is most easily solved *by means of constructing a model*.

In order to solve a problem of the first type within the framework of, say, Newton's theory of perturbations, no more is needed than certain universal laws (in our case Newton's laws of motion) and some of the relevant initial conditions. Initial conditions are, in our case, the masses, velocities, positions, and diameters of the three bodies – sun, earth, and moon – at a certain instant of time (together with the information that only one of these three – the sun – emits light).

In order to consider a problem of the second type, we may construct an actual mechanical model, or refer to a perspective drawing. For our limited purpose, the model may be very rough indeed. It may consist of a fixed lamp, representing the sun, a little wooden earth rotating in a circle round the sun (an ellipse may be too subtle for our rough model), and a little moon rotating in a circle round the earth. One thing would be essential, however: the planes of the two movements must be so inclined towards each other that we obtain lunar eclipses sometimes, but not at *every* full moon.

I call this a 'rough' model because it does not pretend to represent either the actual situation or the actual Newtonian mechanism. It allows neither for the elliptic shapes of the orbits nor for their perturbations. And it may obtain its motion perhaps from a human hand or from a wound-up spring or perhaps from a little electric motor, rather than from Newton's laws of motion. And yet it may serve its purpose very well, since it solves the problem of explanation which has been posed.

A critical discussion of our rough model must give rise, however, to a new problem, 'How are the earth and the moon propelled in the real world?' And with this, we come again to Newton's laws of motion. There is no need, however, to introduce initial conditions into our solution. As far as problems of the second kind are concerned (the explanation of *types* of events), initial conditions may be completely replaced by the construction of a model: this, one might say, incorporates *typical* initial conditions.

So we arrive at the following result.

While explanations or predictions of the first kind – that is, explanations or predictions of singular events – operate with universal laws *and* initial conditions, explanations and predictions of the second kind – that is, those which explain and predict typical events – operate with models, which represent something like typical initial conditions. But the latter also need universal laws if we wish to make the model move, or work, or, as we may say, if we wish to 'animate' the model – that is, if we wish to represent the way in which the various elements of the model may act upon each other.

That these 'animating' laws cannot be dispensed with may be seen if we consider Le Sage's attempt to incorporate the force of attraction into the model of the solar system. Le Sage (and before him Newton) envisaged that 'space' is full of fast particles moving in all directions (think of what we nowadays call 'cosmic radiation') and that the impact of these particles drives the heavy masses towards each other, since each of these masses works like an umbrella in a hailstorm, partially protecting the other masses from the hail. This is an attempt to derive Newton's inverse square law (which otherwise we should have to class as an 'animating' law) from an extension of the model. But even here we need animating laws. We must assume, for example, something like a law according to which at least a proportion of the cosmic particles are absorbed rather than reflected. The same holds true for other attempts to reduce animating laws to structural properties of the model. Such attempts may be highly successful, but they can never reduce *all* the 'animating' laws to 'models' or 'structures'.

Yet the opposite is not the case. It is interesting to see that all specific questions answerable by Newtonian theory may in principle be answered without the construction of a model of the solar system, simply by using the universal laws of motion plus initial conditions. But as a matter of historical fact, models played an all-important role in the development of most theories. It will suffice to remind you that Ptolemy, Copernicus, and Kepler were all makers of models, and that Newton's theory arose partly as an attempt to solve the problem of explaining how Kepler's model was 'animated' – how its elements interacted, and how its moving mechanism worked. In our own century, Rutherford's and Bohr's atom models came many years before quantum mechanics, which provided the (probabilistic) theory of what may be called their 'animation'.

Thus a model consists of certain elements placed in a typical relationship to each other, plus certain universal laws of interaction – the 'animating' laws.

It seems that, as a rule, we first operate with models, and that the models, *together with a rudimentary working mechanism*, may solve a number of problems of the second kind – that is, explain some typical events.

We also see that even in the physical sciences a model need not be a mechanical model. Kepler certainly speculated about the mechanism of his model of the solar system. But while he regarded the model – that is, its elements, and their movements – as fairly well established, he regarded its mode of operation or animation as highly hypothetical, if not as practically unknown. And we must not forget that, although *we* speak of 'Newtonian *mechanics*', Newton himself and his contemporaries looked on action at a distance as something non-mechanical.

Models, as here understood, may be called *'theories'*, or be said to incorporate theories, since they are attempts to solve problems – problems of explanation. But the opposite is far from true. Not all theories are models. Models represent typical initial conditions rather than universal laws. And they therefore need to be supplemented by 'animating' universal laws of interaction – by theories which are not models in the sense here indicated.

All this may be illustrated, for example, by the well-known models of molecules constructed especially by organic chemists. The models of molecules which represent arrangements of atoms may contain sticks representing the chemical bonds. But they do not represent the laws (or resonance) which are the animating laws whereby, we conjecture, the molecules are held together. These laws may, in their turn, be represented by models. But somewhere the model type of theory ends, and the purely abstract animating laws come in which govern the interaction of the various parts or structures that constitute the 'model'.

So much for models in the natural sciences.

Now what about the social sciences? I wish to propose the thesis that what I have said about the significance of models in the natural sciences also holds for models in the social sciences. In fact, models are even more important here because the Newtonian method of explaining and predicting singular events by universal laws and initial conditions is hardly ever applicable in the theoretical social sciences. They operate almost always by the method of constructing *typical* situations or conditions – that is, by the method of constructing models. (This is connected with the fact that in the social sciences there is, in Hayek's terminology, less 'explanation in detail' and more 'explanation in principle' than in the physical sciences.[4])

But the role or function of models in the theoretical social sciences can perhaps be better understood if we look at them from another point of view.

The fundamental problem of both the theoretical and the historical social sciences is *to explain and understand events in terms of human actions and social situations*. The key term here is '*social situation*'.

The description of a concrete historical *social situation* is what corresponds in the social sciences to a statement of *initial conditions* in the natural sciences. And the '*models*' of the theoretical social sciences are essentially descriptions or reconstructions of *typical social situations*.

In my view, the idea of a *social situation* is the fundamental category of the methodology of the social sciences. I should even be inclined to say that almost every problem of explanation in the social sciences requires an analysis of a social situation.

3. An example of situational analysis

Let me explain, with the help of an example, what I call the 'situational analysis of a social situation' or the 'logic of a social situation' or, more briefly, 'situational logic'.

One of my standard examples is a pedestrian, let us call him Richard, who wants to catch a train and is in a hurry to cross a road crowded with moving and parked motor cars and other traffic. Let us assume that what we wish to explain are Richard's somewhat erratic movements in making his way across the road.

What are the obvious situational elements to which we shall have to refer? There are first the various parked motor cars – physical bodies, obstacles, which set certain physical limits to Richard's movements. Then there are moving cars and moving people. They set similar limits to Richard's possible movements, provided we assume that among his many aims there is the aim of avoiding a collision.

But there are further elements in the situation which are equally relevant to the explanation of Richard's movements: the rule of the road, police regulations, traffic signals, zebra crossings, and other such *social institutions*. Some of these social institutions, such as traffic signals or zebra crossings, are linked with, or incorporated in, physical bodies. Others, such as a signalling police constable, are incorporated in human bodies. But others, such as the rule of the road, are of a more abstract nature – yet Richard may experience them exactly as if they were obstacles, either physical bodies such as cars or physical laws (which are 'prohibitions'[5]) such as the law of conservation of momentum pertaining to moving cars. In fact, I propose to use the name 'social institution' for all those things which set limits or create obstacles to our movements and actions almost as if they were physical bodies or obstacles. Social institutions are experienced by us as almost literally forming part of the furniture of our habitat.

But if we wish to explain Richard's movements, then we have to do more than locate the various physical and social obstacles in physical and social space. Indeed, in order that a thing may become an obstacle to Richard's movements, we must first attribute to Richard certain *aims* – the aim, say, of crossing the road in a hurry. Next, we must attribute to him certain elements of knowledge or information – that knowledge, for example, of social institutions which enables him to interpret traffic lights or the signalling of the police constable. (Thus language is a social institution, and so are markets, prices, contracts, and courts of justice.)

Now some social scientists would say that we are operating with psychological assumptions when we attribute to Richard such things as this *information* or these *aims*. But I do not think that is so. A psychologist may even question whether Richard really 'had in mind' anything like an 'aim' of crossing the road or whether, rather, his only 'aim' in a psychological sense, was to avoid missing his train, and whether he was not entirely absorbed by this one idea. Subsidiary aims, such as crossing the road, or putting one foot before the other, or keeping his balance while walking, or holding on to his attaché case, may all be non-existent, psychologically speaking, even though we may by logical analysis recognize them as intermediate aims which, under the given conditions, are prerequisites for achieving the ultimate aim of catching the train.

However this may be, I propose to treat both Richard's aims and Richard's knowledge not as psychological facts, to be ascertained by psychological methods, but as *elements of the objective social situation*. And I propose to treat his actual psychological aim of catching a train as irrelevant for solving our particular problem, which only requires that his aim – his 'situational aim' – was to cross the road as quickly as was compatible with safety. Similarly, we shall not be interested in Richard's knowledge in general – his familiarity with Verdi's operas, say, or with certain Sanskrit texts – even though a psychological enquiry could have brought out how great a part Verdi and Sanskrit play in his thoughts, how at the very moment of his crossing the road he may have hummed a passage from Verdi, or considered the adequacy of a translation of a passage from the *Atharva-veda*. We shall be concerned only with information or knowledge (such as his knowledge of the rule of the road) which is relevant to the situation.[6]

Thus the situational analysis will comprise some physical things and some of their properties and states, some social institutions and some of their properties, some aims, and some elements of knowledge. Given this analysis of the social situation, we may be able to explain, or to predict, Richard's movements as he crosses the street.

Clearly, what we have here is a model, a typical case rather than a singular case. Even though our problem may change, and we may some day be interested in explaining a singular event (say, how and why Richard was held up by the traffic on a certain day so that he did not catch his train, and therefore missed a great performance of Verdi's *Otello*, or an interesting meeting of the Buddhist society), our method of situational analysis always turns Richard into 'anybody' who may share the relevant situation, and reduces his personal living aims and his personal knowledge to elements of a *typical situational model*, capable of 'explaining in principle' (to use Hayek's term) a vast class of structurally similar events.

My thesis is that only in this way can we explain, and understand, a social event (*only* in this way because we never have sufficient laws and initial conditions at our disposal to explain it with their help).[7]

Now if situational analysis presents us with a model, the question arises: what corresponds here to Newton's universal laws of motion which, as we have said, animate the model of the solar system? Or in other words, how is the model of a social situation animated?

4. Psychologism

The usual mistake here is to assume that in the case of human society, the animation of a social model has to be provided by the human *anima* or *psyche*, and that here, therefore, we have to replace

Newton's laws of motion either by the laws of human psychology in general, or perhaps by the laws of individual psychology pertaining to the individual characters who are involved as actors in our situation.

But this is a mistake, for several reasons. First of all, we have already *replaced* Richard's concrete, conscious or unconscious, psychological experiences by some abstract and typical situational elements, such as those we dubbed 'aims' and 'knowledge'. Secondly, it is the central point of situational analysis that we need, in order to 'animate' it, no more than the assumption that the various persons or agents involved act *adequately*, or *appropriately* – that is to say, in accordance with the situation. We must remember, of course, that the situation, as I use this term, already contains all the relevant aims and all the available relevant knowledge, especially of the various possible means for realizing these aims.

Thus there is only one animating law involved – the principle of acting appropriately to the situation, which is clearly an *almost empty* principle. It is known in the literature under the name 'rationality principle', a name which has led to countless misunderstandings.

If you look upon this so-called rationality principle from the point of view which I have here adopted, then you will find that it has little or nothing to do with the empirical or psychological assertion that man always, or in the main, or in most cases, acts rationally. Rather, it turns out to be an aspect of, or a consequence of, the methodological postulate that we should pack or cram our whole theoretical effort, our whole explanatory theory, into an analysis of the *situation* – into the *model*.

If we adopt this methodological postulate, then the animating law will, as a consequence, become a kind of zero-principle.[8] For the principle may be stated as follows: once we have constructed our model of the situation, we assume no more than that the actors act within the terms of the model, or that they 'work out' what was *implicit* in the situation. This, incidentally, is what the term 'situational logic' is meant to allude to.

The adoption of the rationality principle can therefore be regarded as a by-product of a methodological postulate. It does not play the role of an empirical explanatory theory, of a testable hypothesis. For in this field, the empirical explanatory theories or hypotheses are, rather, our various models, our various situational analyses. It is these which may be empirically more or less adequate – which may be discussed and criticized, and whose adequacy may sometimes even be tested, thereby enabling us, if they fail the test, to learn from our mistakes.

Tests of a model, it has to be admitted, are not easily obtainable and are usually not very clear-cut. But this difficulty arises even in the physical sciences. It is connected, of course, with the fact that models are always and necessarily somewhat rough and schematic over-simplifications. Their roughness entails a comparatively low degree of testability. For it will be difficult to decide whether a discrepancy is due to the unavoidable roughness, or to a mistake in the model. Nevertheless, we can sometimes decide, by testing, which of two competing models is the better. And in the social sciences, tests of a situational analysis can sometimes be provided by historical research.

5. Further examples

I have treated the example of Richard trying to cross the road in some detail, because I believe that it contains almost all the relevant elements of situational analysis as used in economics, in social anthropology, in the sociology of power politics, and in social or political history.

To take a familiar example, the most important part of classical economic theory is the theory of perfect competition. It may be developed as the situational logic of an idealized or over-simplified social situation – the situation of people acting within the institutional framework of a perfectly free market in which buyers and sellers are equally informed of the physical qualities of the goods which are bought and sold. Similarly, the pure theory of monopoly or of duopoly is nothing but the situational logic of certain idealized social situations.

Analogous remarks may be made, for example, about social anthropology. Social anthropology tries (or should try) to describe the institutional and traditional framework as well as the problems of a society in such a way that the typical actions of its members become rationally understandable as appropriate. It also tries to explain, in part, the institutional framework itself, and its changes, as the result (as a rule, the unintended result) of actions taken in certain historical situations, such as, for example, the clash of two cultures. (That much of the astonishing development of ancient Greece is influenced by culture clash may be seen, for example, from the scanty fragments that remain of the work of Hecataeus of Abdera.[9])

6. Problem situations

What I have said so far is a brief outline of the methodology of the explanatory social sciences, especially economic theory and social anthropology. But it applies especially to historical explanations, which always operate with a rational reconstruction of a situation.[10]

Perhaps the best field for this methodology is the history of science. Here the situation of the agent – the creative scientist – is the problem situation which he finds in his scientific field, though he may, of course, reframe it by looking at it in a new way. One may generalize this and say that *whenever we wish to explain or understand history, we have to look at it as a history of problem situations.*

7. Instruments and truth: the falsity of social theories

I now turn to the last part of my lecture.[11] In this, I intend first to develop certain arguments which favour pragmatism, but afterwards I will explain why I do not agree with pragmatism, and that I regard theories as steps towards the truth.

You will remember my assertion that the rationality principle does not play the role of an empirical or psychological proposition, and, more especially, that it is not treated in the social sciences as subject to any kind of test. Tests, when available, are used to test a particular model, a particular situational analysis – but not the general method of situational analysis, and not, for this reason, the rationality principle: to uphold this is part of the method. (The general method is not testable, though it is arguable. The main argument in favour of it is that it appears to give rise to better testable explanatory hypotheses – that is, conjectural situational models – than other methods.[12]) Thus, if a test indicates that a certain model is less adequate than another one, then, since both operate with the rationality principle, we have no occasion to discard this principle.

This remark explains, I think, why the rationality principle has frequently been declared to be *a priori*. And indeed, what else could it be if it is not empirical?

This point is of considerable interest. Those who say that the rationality principle is *a priori* mean, of course, that it is *a priori* valid, or *a priori* true. But it seems to me quite clear that they must be wrong. For the rationality principle seems to me clearly false – even in its weakest zero[13] formulation, which may be put like this: 'Agents always act in a manner appropriate to the situation in which they find themselves.'

I think one can see very easily that this is not so. One has only to observe flustered drivers trying to get out of a traffic jam, or desperately trying to park their cars when there is hardly any parking space to be found, or none at all, in order to see that we do not always act in accordance with the rationality principle. (In thus hoping against hope, we do not act rationally, even though we act in accordance with a psychological mechanism whose evolution is rationally understandable.[14]) Moreover, there are, obviously, vast personal differences, not only in knowledge and skill – these are part of the situation – but also in assessing or understanding a situation. And this means that some people will act appropriately and others not.

But a principle that is not universally true is false. Thus the rationality principle is false. I think there is no way out of this. Consequently, we must deny that it is *a priori* valid.

Now if the rationality principle is false, then an explanation which consists of the conjunction of this principle and a model must also be false, even if the particular model in question is true.

But can the model be true? Can any model be true? I do not think so. Any model, whether in physics or in the social sciences, must be an over-simplification. It must omit much, and it must over-emphasize much.

Take a Newtonian model of the solar system. Even if we assume that Newton's laws of motion are true, the model would not be true. Though it contains a number of planets – in the form, incidentally, of mass-points, which they are not – it contains neither the meteorites nor the cosmic dust. It contains neither the pressure of the light of the sun nor that of cosmic radiation. It does not even contain the magnetic properties of the planets, or the electrical fields which result in their neighbourhood from the movement of these magnets. And – perhaps most important – it does not contain anything representing the action of the distant masses upon the bodies of the solar system. It is, like all models, a vast over-simplification.

I think we have to admit that most successful scientific theories are lucky over-simplifications. But although this does not necessarily impugn the veracity of universal laws, it seems to be quite unavoidable in the construction of models, both in the natural and in the social sciences, that they over-simplify the facts, and thus do not represent the facts truly.

Now if the rationality principle, which in the social sciences plays a role somewhat analogous to the universal laws of the natural sciences, is false, and if in addition the situational models are also false, then both the constituent elements of social theory are false. If we wish nevertheless to uphold the method of situational analysis as *the* proper method of the social sciences, as I certainly do, and if we wish to uphold the view that science searches for truth, are we not in a hopelessly difficult position?

8. Instrumentalism

There is a group of philosophers who may (somewhat precipitately) be pleased about what I have been saying: the pragmatists or instrumentalists. For it is their creed that we should not, or cannot, aim at 'pure' knowledge – or at truth – with our scientific theories: that scientific theories are *nothing but instruments* – instruments, that is, for prediction or practical application – and that we deceive ourselves if we think that theories may afford us either explanations or an understanding of what actually happens in the world.

Thus the instrumentalists might well rejoice, for all that I have said seems to support their view. And they may even point out that the difficulties which I have mentioned are an old story, and that at least since Niels Bohr, instrumentalism has been almost universally accepted by physicists.

Now I have to admit that, owing to the authority of Niels Bohr, instrumentalism has become very fashionable among physicists. But the list of those who resisted the lure of this fashion includes Einstein, de Broglie, and Schrödinger. This gives me the courage to confess that I too am an anti-instrumentalist (or, as I may perhaps say, a realist). In fact I have elsewhere combated the creed of instrumentalism at some length,[15] though I have so far criticized instrumentalism only as a philosophy of the *physical* sciences.

What do we anti-instrumentalists assert? We admit, of course, that a scientific theory may be applied to all sorts of practical problems, either at once, when first invented, or at a later date. And we therefore have no objection to the assertion that all scientific theories are instruments – either actual or potential instruments. But we assert that they are not *merely* instruments. For we assert that we may learn from science something about the structure of our world: that scientific theories can offer genuinely satisfying explanations that can be understood and so add to our understanding of the world. And we assert – this is the crucially important point – that science aims at truth, or at getting nearer to the truth, however difficult it may be to approach truth, even with very moderate success.

The issue can be put in a nutshell: are scientific theories *nothing but* instruments, or should they, as I suggest, be regarded as attempts to find the truth about our world, or at least as attempts to get nearer to the truth?

But, you may ask, is it permissible to speak so glibly about the truth? And even if it is, is it permissible to speak so glibly about approaching truth or getting nearer to the truth? Are not all these words simply without meaning?

These are important objections. Let me first deal with the question of the meaningfulness of the word 'truth'.

9. Truth

It is strange that so many people should believe that there is no answer to Pilate's question 'What is truth?' For, after all, in thousands of courts of justice thousands of witnesses are admonished every day to speak the truth. And most of them seem to know very well what is expected from them.

In fact, there is an old answer to the question: 'When is an assertion, or a proposition, or a statement, or a theory, or a belief true?' The answer is: *An assertion is true if it corresponds to, or agrees with, the facts.*

But what does it mean to say that an assertion or a theory corresponds to the facts? This question, too, has been satisfactorily answered – by the mathematician and logician Alfred Tarski.

I cannot, of course, give a full account of Tarski's theory here.[16] Suffice it to say that it is a theory which is in complete accord with the common-sense view that the statements 'snow is white' and 'grass is green' are true, while the statements 'snow is green' and 'grass is red' are false.

Tarski's theory shows that we are fully entitled to use, without any qualms, the words 'true' and 'false' in their ordinary senses. It also shows that there cannot be, in any language comparable in its wealth of expressions to our ordinary European languages, a *general criterion of truth* – that is to say, a general method by which we could decide, of any given proposition, whether or not it is true.

Thus we are not, as a rule, able to decide of a statement or a theory whether it is true. Ascertaining truth may be a very difficult business, and often a practically impossible one. But the meaningfulness of the term 'truth' is affected by this no more than the meaningfulness of the term 'father' is affected by any difficulty in ascertaining fatherhood.

If we eliminate from language ambiguous terms like 'yesterday', a term which today means something different from what it will mean tomorrow, and if we take some further similar precautions, then it follows from Tarski's theory that every statement in this purified language will be either true or false, with no third possibility. Moreover, we can have an operation of negation in our language such that if a proposition is not true, then its negation is true.

This shows that of all propositions one half will be true and the other half false. So we can be sure that there will be lots of true propositions, even though we may have great trouble in finding out which they are.

10. Approaching truth

I now turn to the second question – whether we can meaningfully speak about *approaching truth*, or *getting nearer to the truth*, or, more precisely, whether it can be meaningful to say of a theory that it is a better approximation to the truth than another theory.

I worked on this question for a considerable time before I could produce an answer. But with the help of Tarski's concept of truth, and a few other purely logical concepts (especially the concept of logical content, also due to Tarski), I think that I have been able to give a purely logical definition of the relation '*a* is a better approximation to the truth than *b*', or '*a* is more similar to the truth than *b*'. This definition (which may be found in my *Conjectures and Refutations*[17]) is, like most definitions, of little significance in itself. What is significant is that it establishes one thing: that the much suspected phrase '*a* is a better approximation to the truth than *b*' is certainly not meaningless.

There are many examples in physics of competing theories which form a sequence of theories such that the later ones appear to be better and better approximations to the (unknown) truth.

For example, Copernicus' model appears to be a better approximation to the truth than Ptolemy's, Kepler's a better approximation than Copernicus', Newton's theory a better approximation still, and Einstein's better again.

It is very interesting in this connection to note that Einstein did not present his theory of gravitation as a true theory. On the contrary, he argued that it could not be true, and he spent more than thirty years of his life trying to improve his own theory. But in spite of all this, he always believed that it was a better approximation to the truth than Newton's theory and other theories (such as Milne's).

11. Reply to instrumentalism

I shall now conclude my lecture with a reply to instrumentalism. I will be very brief, and I will confine myself to the problem raised by *the known falsity of social theories*.

I think that I am now in a position to answer those instrumentalists who earlier in this lecture may have welcomed my description of the methods of the social sciences as a confirmation of their philosophy of science.

My answer is this: if my view of the social sciences and their methods is correct, then, admittedly, no explanatory theory in the social sciences can be expected to be true. *Nevertheless, this need not trouble an anti-instrumentalist.* For he may be able to show that those methods may be very good methods, in the sense that they make it possible for us to discuss critically *which of the competing theories, or models, is a better approximation to the truth*.

This, I suggest, is the situation in the social sciences. In seeking pure knowledge our aim is, quite simply, to understand – to answer how-questions and why-questions (but *not* pseudo-questions of the 'what is ...?' form). These are questions which are answered by giving an explanation. Thus all problems of pure knowledge are *theoretical problems*: they are *problems of explanation*.

A problem of this kind may well have originated in a *practical* one. For example, a *practical problem* such as 'What can be done to combat poverty?' has led to the purely theoretical problem 'Why are people poor?' and from there to the theory of wages and prices, and so on – in other words, to pure economic theory, which of course constantly creates its own new theoretical problems. In the development of the theory, the problems dealt with – and especially the unsolved problems – multiply, and they become differentiated, as they always do when our knowledge grows.

12. Rationality and the status of the rationality principle[18]

My views on the rationality principle have been closely questioned. I have been asked whether there is not some confusion in what I say about the status of the 'principle of acting adequately to the situation' (that is, of my own version of the 'rationality principle'). I was told, quite rightly, that I should make up my mind whether I want it to be a methodological principle, or an empirical conjecture. If it were a methodological principle, it would be clear why it could not be empirically tested and why it could not be empirically false (but only part of a successful or unsuccessful methodology). If it were an empirical conjecture, it would become part of the various social theories – the 'animating part' of every social model. But then it would have to be part of some empirical theory, and would have to be tested along with the rest of that theory, and rejected if found wanting.

This second case is the one that corresponds better to my own view of the status of the rationality principle: I regard the principle of adequacy of action (that is, the rationality principle) as an integral part of every, or nearly every, testable social theory.

Now if a theory is tested, and found faulty, then we have always to decide which of its various constituent parts we shall make accountable for its failure. My thesis is that it is *sound methodological policy* to decide not to make the rationality principle, but the rest of the theory – that is, the model – accountable.

In this way it may appear that in our search for better theories we treat the rationality principle as if it were a logical or a metaphysical principle exempt from refutation: as unfalsifiable, or as *a priori* valid. But this appearance is misleading. There are, as I have indicated, good reasons to believe that the rationality principle, even in my minimum formulation, is actually false, though a good approximation to truth. Thus it cannot be said that I treat it as *a priori* valid.

I hold, however, that it is good policy, a good methodological device, to refrain from blaming the rationality principle for the breakdown of our theory. For we learn more if we blame our situational model. The policy of upholding the principle can thus be regarded as part of our methodology.

The main argument in favour of this policy is that our model is far more interesting and informative, and far better testable, than the principle of the adequacy of our actions. We do not learn much in learning that this is not strictly true: we know this already. Moreover, in spite of being false, it is as a rule sufficiently near to the truth: if we can refute our theory empirically, then its breakdown will, as a rule, be pretty drastic, and though the falsity of the rationality principle may be a contributing factor, the main responsibility will normally attach to the model. Also, the attempt to replace the rationality principle by another one seems to lead to complete arbitrariness in our model-building. And lastly, we must not forget that we can test a theory only as a whole, and that the test consists in finding the better of two competing theories which may have much in common, and that most of them have the principle of adequacy in common.

13. 'Irrational' actions

But suppose that we are interested in a certain action not as an approximation to, but as a deviation from, the action prescribed by the logic of the situation as I have so far discussed it. Suppose that our problem is to understand the actions of a person who acts inappropriately to his situation.[19]

Churchill said in *The World Crisis*[20] that wars are not won but only lost – that, in effect, they are competitions in incompetence. Does not this remark provide us with a kind of model for typical social and historical situations: *a model not animated by the rationality principle of the adequacy of our actions, but by a principle of inadequacy*?

The answer is that Churchill's dictum means that most war leaders are inadequate to their task, rather than that their actions cannot be understood (in good approximation, at least) as adequate for the situation *as they see it*.

In order to understand their (more or less inadequate) actions, we have therefore to reconstruct a wider view of the situation than their own. This must be done in such a way that we can see how and why the situation as they saw it (with their limited experience, their limited or overblown aims, their limited or overexcited imagination) led them to act as they did – that is to say, adequately for their inadequate view of the situational structure. Churchill himself uses this method of interpretation with great success, for example in his careful analysis of the failure of the Auchinleck–Ritchie team (in volume IV of *The Second World War*).[21]

It is interesting to me that we employ the rationality principle to the limit of what is possible whenever we try to understand the action of a madman. We try to explain a madman's actions, as far as possible, by his aims (which may be monomaniacal) and by the 'information' on which he acts – that is to say, by his convictions (which may be obsessions, that is, false theories so tenaciously held that they become practically incorrigible). In so explaining the actions of a madman we explain them in terms of our wider knowledge of a problem situation which comprises his own, narrower, view of his problem situation. And understanding his actions means seeing their adequacy according to his view – his madly mistaken view – of the problem situation.

We may in this way even try to explain how he arrived at his madly mistaken view: how certain experiences shattered his originally sane view of the world and led him to adopt another – the most rational view he could develop in accordance with the information at his disposal – and how he had to make this new view *incorrigible*, precisely because it would break down at once under the pressure of refuting instances, which would leave him (so far as he could see) stranded without any interpretation of his world – a situation to be avoided at all costs, from a rational point of view, since it would make all rational action impossible.

Freud has often been described as the discoverer of human irrationality. But this is a misinterpretation, and a very superficial one to boot. Freud's theory of the typical origin of a neurosis falls entirely into our schema of explanations incorporating both a situational model *and* the rationality principle. For he explains a neurosis as an attitude adopted in early childhood because it was the best available way out of a situation which the child was unable to understand and cope with. Thus the adoption of the neurosis becomes a rational act of the child – as rational, say, as the act of a man who, jumping back when confronted by the danger of being run over by a car, collides with a bicyclist. It is rational in the sense that the child chose what appeared to him the immediate, or the obvious, or perhaps the lesser evil – the less intolerable of two possibilities.

I shall say no more here about Freud's method of therapy than that it is even more rationalistic than his method of diagnosis or explanation (for it is based on the assumption that once a man fully understands what befell him as a child, his neurosis will pass away).

But if we thus explain everything in terms of the rationality principle, does it not become tautological? By no means. For a tautology is obviously true, whilst we make use of the rationality principle merely as a good approximation to the truth, recognizing that it is not true.

But if this is so, what becomes of the distinction between rationality and irrationality? Between mental health and mental disease?

This is an important question. The main distinction, I suggest, is that a healthy person's beliefs are not incorrigible: a healthy person shows a certain readiness to correct his beliefs. He may do so only reluctantly, yet he is nevertheless ready to correct his views under the pressure of events, of the opinions held by others, and of critical arguments.

If this is so then we can say that the mentality of the man with definitely fixed views, the 'committed' man, is akin to that of the madman. It may be that all his fixed opinions are 'adequate' in the sense that they happen to coincide with the best opinion available at the time. But in so far as he is committed, he is not rational: he will resist any change, any correction. And since he cannot be in the possession of the precise truth (*nobody is*) he will resist rational correction of even wildly mistaken beliefs. And he will resist even if their correction is widely accepted during his lifetime.

Thus when those who praise commitment and irrational faith describe themselves as irrationalists (or post-rationalists) I agree with them. *They are irrationalists*, even if they are capable of reasoning. For they take pride in rendering themselves incapable of breaking out of their shell: in making themselves prisoners of their manias. They make themselves spiritually unfree, by an action whose adoption we may explain (following the psychiatrists) as one that is rationally understandable: understandable, for example, as an action which they commit owing to fear – fear of being compelled, by criticism, to surrender a view which they dare not give up since they make it (or believe they make it) the basis of their whole life. ('Free commitment' and fanaticism – which, we know, can border on madness – are thus related in the most dangerous manner.)

To sum up: we should distinguish between rationality as a personal attitude (which, in principle, all sane men are capable of sharing) and the rationality principle.

Rationality as a personal attitude is the attitude of readiness to correct one's beliefs. In its intellectually most highly developed form it is the readiness to discuss one's beliefs critically, and to correct them in the light of critical discussions with other people.

The 'rationality principle', on the other hand, has nothing to do with the assumption that men are rational in this sense – that they always adopt a rational attitude. It is, rather, a minimum principle (since it assumes no more than the adequacy of our actions to our problem situations *as we see them*) which animates almost all our explanatory situational models, and which, although we know that it is not true, we have some reason to regard as a good approximation to the truth. Its adoption reduces considerably the arbitrariness of our models, an arbitrariness which becomes capricious indeed if we try to do without this principle.

NOTES

1 I was particularly impressed by Hayek's formulation that economics is the 'logic of choice'. (Cp., for example, F.A. von Hayek, 'Economics and Knowledge' (1936) reprinted in *Individualism and Economic Order*, Routledge & Kegan Paul and the University of Chicago Press, Chicago, 1948; see p. 35.) This led me to my formulation of the 'logic of the situation' (cp. my *Poverty of Historicism*, p. 149). This seemed to me to embrace, for example, the logic of choice and the logic of historical problem situations. (The origin of this idea may explain why I rarely stressed the fact that I did not look at the logic of the situation as a deterministic theory: I had in mind the logic of situational choices.)

2 Sections 12 and 13, added after the original lecture (see note 19), contain a further discussion of the 'rationality principle'.

3 Here the word 'only' is intended to stress my opposition to that empiricist tradition which looks upon science as being based on observations and experiments. Of course, this passage needs to be amplified – for example, by a discussion of corroboration. Cp. appendix *9 of my *Logic of Scientific Discovery* and chapter 10 of my *Conjectures and Refutations*.

4 See 'Degrees of Explanation', in F.A. von Hayek, *Studies in Philosophy, Politics and Economics*, Routledge & Kegan Paul, London, 1967, pp. 3–21.

5 Physical laws (say, that of conservation of energy) 'prohibit' the occurrence of certain things (say, the construction of a perpetual motion machine). Cp. my *Logic of Scientific Discovery*, p. 69 and my *Poverty of Historicism*, p. 61.

6 It seems to me that we clarify the nature of social theory if, as is suggested in the text, we de-psychologize the aims, information, and knowledge of the actors in typical social situations. (Note that this is not a concession to behaviourism.)

Consider, for example, the old controversy in economics about profit maximization. According to the theory of profit maximization, the businessman maximizes his (monetary) profits by a policy of marginal cost pricing. In a paper by R.L. Hall and C.J. Hitch ('Price Theory and Business Behaviour', *Oxford Economic Papers*, 2, 1939, pp. 12–45), however, this theory was criticized on the basis of empirical evidence, obtained by means of questionnaires, about the way in which business-men decided their pricing policy. It has been suggested that this evidence shows the theory of profit maximization to be false. This led defenders of the theory to suggest that it was not meant to describe the actual behaviour of businessmen, but that it was only a tool, or an *instrument*, for prediction. Thus both groups seem to have been in agreement in assuming that the profit maximization theory psychologized the aims and the information of the agents (businessmen) in the typical social situation considered.

Against all this, I suggest that the method of situational logic is not concerned with what the agent's actual thoughts were when performing the action (compare the case of Richard's crossing the street). In consequence, evidence from questionnaires about psychological mo-tivation is not necessarily relevant to the testing of a theory about situational logic.

As to the status of the profit maximization model (from the point of view here discussed), not only could one admit its falsity as a model of businessmen's *psychological motivation* without being forced into instrumentalism, but even if it is *false* as a theory of businessmen's behaviour, it may be still valued as an approximation to the truth.

I should not object to alternative situational models in which, say, businessmen's behaviour is explained in terms of an aim to increase the standing of their firm or even their own position within it. In such a model, the maximization of profits might enter, not as an aim, but as the result of a kind of *situational constraint*. (Cp. Adam Smith, *The Wealth of Nations*, book I, chapter XI, part I: '... good management, ... can never be universally established but in consequence of that free and universal competition which *forces* everybody to have recourse to it for the sake of self-defence'. The italics are mine. I owe this reference to Jeremy Shearmur.) It is doubtful whether the difference between such a model and the view that takes the maximization of profits as an aim is worth bothering about, though this will, of course, depend on what we want to explain: on what we regard as our problem.

7 The material in parentheses was added in 1974.

8 Cp. my *Poverty of Historicism*, pp. 141f., for a discussion of the 'zero method'.

9 Cp. H. Diels and W. Kranz, *Die Fragmente der Vorsokratiker*, 6th edition, Weidmannsche Verlagsbuchhandlung, Berlin-Grunewald, 1951–2, volume II, pp. 240–4.

10 See my 'On the Theory of the Objective Mind', chapter 4 in my *Objective Knowledge*, especially sections 9–12. Cp., particularly, the discussion of R.G. Collingwood in section 12.

11 That is, of the original lecture.

12 The sentence in parentheses was added in 1974. The problem of the status of the rationality principle is more fully discussed in section 12.

13 Cp. the reference given in note 8.

14 The sentence in parentheses was added in 1974.

15 See my *Conjectures and Refutations*, chapter 3.

16 I have discussed it in my *Conjectures and Refutations* – see, for example, pp. 223–7 – and in my *Objective Knowledge* – see especially pp. 308–18 and chapter 9.

17 See my *Conjectures and Refutations*, chapter 10 and addendum 3 (and also my *Objective Knowledge*, chapters 2 and 9). For criticism of my definition, see David Miller, 'The Truth-likeness of Truthlikeness', *Analysis, 33*, 1972, pp. 50–5. See also, in *The British Journal for the Philosophy of Science, 25*, 1974: David Miller, 'Popper's Qualitative Theory of Verisimilitude', pp. 166–77; David Miller, 'On the Comparison of False Theories by their Bases', pp. 178–88; and Pavel Tichý, 'On Popper's Definitions of Verisimilitude', pp. 155–60.

18 This section, and the final one (which formed part of it) were added after the lecture, on the basis of the discussion following the lecture.

19 In the earlier part of the lecture, I was concerned with the rationality principle considered as the view that people act appropriately to the objective situation in which they find themselves (including their knowledge and skills). In what follows, I am concerned with the view that people act in a manner appropriate to the situation as they see it.

It seems to me now that there are at least three senses of 'rationality' (and, accordingly, of the 'rationality principle'), all objective, yet differing with regard to the objectivity of the situation in which the agent is acting: (1) *The situation as it actually was* – the objective situation which the historian tries to reconstruct. Part of this objective situation is (2) *The situation as the agent actually saw it*. But I suggest that there is a third sense intermediate between (1) and (2): (3) *The situation as the agent could* (within the objective situation) *have seen it*, and perhaps ought to have seen it. It is clear that there will be three senses of the 'rationality principle' corresponding to these three senses of 'the situation'. It is further clear that the difference between (1) and the two other versions of the rationality principle will play a part in our understanding of action, especially in the historian's attempt to explain *failure*, and that the difference between (2) and (3) will play a similar part. (2) and (3), it should be stressed, form in their turn part of (a more or less elaborate analysis of) the objective situation (1). Moreover, if there is a clash between (2) and (3), then we may well say that the agent did not act rationally. (Such a clash would, I think, be described by psychoanalysts as a failure of the reality principle.) (3) may well include an appreciation of the difficulties in seeing certain aspects of the situation as they were.

With regard to the paper, I could be said, using these formulations, to have worked with (1) and (3) in the earlier sections, and with (2) in the final sections. I might add that, in my view, we sometimes act in a manner not adequate to the situation in any of the senses (1), (2), or (3) – in other words, that the rationality principle is not universally true as a description of our ways of acting.

20   Winston Churchill, *The World Crisis*, Thornton Butterworth, London, 1923–31.

21   Winston Churchill, *The Second World War*, volume IV: *The Hinge of Fate*, Cassell & Co., London, 1951. See chapter 26.

9

EPISTEMOLOGY AND INDUSTRIALIZATION

Francis Bacon looked forward to an alteration in the form of production and to the effective control of nature by man, as a result of a change in the ways of thinking.

Karl Marx

I

In a famous and highly dramatic passage of his chief work, Plato demands that philosophers should be kings and, *vice versa*, that kings – or autocratic rulers – should be fully trained philosophers.[1] Plato's proposal that philosophers should be kings has pleased many philosophers, and some of them have taken it quite seriously. Personally I do not find it an attractive proposal. Quite apart from the fact that I am against any form of autocracy or dictatorship, including the dictatorship of the wisest and best, philosophers do not seem to me particularly well suited for the job. Take for example the case of Thomas Masaryk, the creator, first president and, one might say, the philosopher-king of the Czechoslovak Republic. Masaryk was not only a fully trained philosopher, but also a born

A lecture delivered in German on 13 June 1959, at the School of Economics, the University of St Gallen, Switzerland, in a series of lectures entitled 'Europe: Inheritance and Future Tasks', on the topic 'The Influence of Philosophy upon some Fundamental Turning Points in the History of Europe'. First published, with revisions, in *Ordo, 30*, (*Festschrift* for F.A. von Hayek), Gustav Fischer Verlag, Stuttgart and New York, 1979. Jeremy Shearmur has provided the references in the notes. The motto is taken from Karl Marx, *Capital*, volume I, chapter 13, section 2 (the footnote on pp. 413ff of the Everyman's Library edition, J.M. Dent & Sons Ltd., London, E.P. Dutton & Co. Inc., New York, 1930 and subsequent impressions. Cp. Lawrence & Wishart, London/Progress Publishers, Moscow edition, 1963 and subsequent impressions, footnote 2 on p. 368).

statesman and a great and admirable man. And his creation, the Czechoslovak Republic, was an unparalleled political achievement. Yet the dissolution of the Old Austrian Empire was also partly Masaryk's work. And this proved a disaster for Europe and the world. For the instability that followed this dissolution was largely responsible for the rise of Nazism and finally even for the downfall of Masaryk's own Czechoslovak Republic. And it is significant that Masaryk's doctrine that 'Austria-Hungary, this antinational ... state, must be dismembered'[2] (to use his own words) was derived from a mistaken *philosophical doctrine*: from *the philosophical principle of the National State.*[3] But this principle, the principle of political nationalism, is not only an unfortunate and even a mischievous conception, it is also one that is actually impossible to realize. This is because Nations – in the sense of those who advocate this principle – do not exist: they are theoretical constructs, and the theories in which they are constructed are wholly inadequate, and completely inapplicable to Europe. For the political theory of nationalism rests on the assumption that there are ethnic groups which at the same time are also linguistic groups, and which happen also to inhabit geographically unified and coherent regions with natural boundaries that are defensible from a military point of view – groups which are united by a common language, a common territory, a common history, a common culture, and a common fate. The boundaries of the regions inhabited by these groups should, according to the theory of the Nation State, form boundaries of the new national states.

It was this theory which underlay the Masaryk–Wilson principle of the 'Self-determination of Nations', and in its name the multilingual state of Austria was destroyed.

But no such regions exist – at least not in Europe nor, indeed, anywhere in the Old World.[4] There are few geographical regions in which only one native language is spoken: almost every region has its linguistic or 'racial' minority. Even Masaryk's own newly created national state contained, in spite of its small size, several linguistic minorities.[5] And the principle of the National State played a decisive role in its destruction: it was this principle which allowed Hitler to appear in the role of a liberator, and which confused the West.

It is important to my present theme that the idea of nationalism is a *philosophical* idea. It sprang from the theory of sovereignty – the theory that power in the state must be undivided – and from the idea of a superhuman ruler who rules by the grace of God. Rousseau's replacement of the king by the people only inverted the view: he made of the people a nation – a superhuman nation by the grace of God. Thus the theory of political nationalism originated in a philosophical inversion of the theory of absolute monarchy. The history of this development seems to me characteristic of the rise of many philosophical ideas, and it suggests to my mind the lesson that philosophical ideas should be treated with a certain reserve. It also may teach us that there are fundamental ideas, such as the idea of political liberty, of the protection of linguistic and religious minorities, and of democracy, which remain fundamental and true even when defended by untenable philosophical theories.

The fact that an admirable man and a great statesman like Masaryk was led by certain philosophical ideas to commit so grave a blunder – that he accepted a philosophical theory which was not only untenable but, under existing conditions, almost bound to destroy his work as a statesman – all this, I believe, amounts to a strong argument against Plato's demand that philosophers should rule. But one could also adduce another and entirely different argument against Plato: one could also say that Plato's demand is superfluous, because the philosophers are ruling anyway – not officially, it is true, but all the more so in actual fact. For I want to advance the thesis that the world is ruled by ideas: ideas both good and bad. It is, therefore, ruled by those who produce those ideas – that is, by the philosophers, though rarely by professional philosophers.

The thesis that it is the philosophers who actually are the rulers is, of course, not new. Heinrich Heine expressed it as follows in 1838:[6]

> Mark this, ye proud men of action: Ye are nothing but unconscious instruments of the men of thought who, often in humblest seclusion, have appointed you to your favourite task. Maximilien Robespierre was merely the hand of Jean Jacques Rousseau ...

And Friedrich von Hayek, in his great work of liberal political philosophy, *The Constitution of Liberty*, has emphasized the relevance of this idea for us today, and its importance in the liberal tradition.[7]

Countless examples illustrate the political power of philosophical ideas. Marxism is a philosophy: Marx himself proudly quoted a review which described his theory as expounded in *Capital*, quite correctly, as the last of the great post-Kantian philosophical systems.[8] His seizure of power, thirty-four years after his death, in the person of Lenin, is an almost exact repetition of the seizure of power by Rousseau, sixteen years after his death, in the person of Robespierre.

Orthodox Marxism, of course, denies the thesis of the political power of ideas: it sees in ideas mainly the inevitable consequences of technical and industrial developments. What changes first, Marx teaches, is the technique of production. Depending on this, the class-structure of society will change, followed by the prevalent ideas. And finally, when the whole substructure has changed, the system of political power will also change.[9] But this theory – which contradicts our thesis of the power of philosophical ideas – is refuted by history. Consider, for example, the history of Russia since 1917. What came there first was the seizure of power – that is, what according to Marx's theory should have come last. Next came Lenin's great *idea*: the idea that socialism is the dictatorship of the proletariat plus electrification.[10] And last came electrification, industrialization, and the enforced change of the so-called economic 'substructure'. This change was thus imposed from above, by a new instrument of power, the new class dictatorship.

Later I shall endeavour to show that the first industrial revolution – the English Industrial Revolution – was also inspired by philosophical ideas.

A totally different example of the seizure of political power by a philosophy – a seizure of power by purely democratic means – has been brought to my attention by Hayek. The English philosopher and economist, John Stuart Mill, wrote in his *Autobiography*, published shortly after his death in 1873, that in the years around 1830 his circle (the so-called Philosophical Radicals) had adopted the following programme: they wanted to achieve an improvement in human society by 'securing full employment at high wages to the whole labouring population'.[11] Seventy-two years after his death John Stuart Mill seized political power in England. And no political party would dare today to challenge his programme.

II

The political power of philosophical ideas – and often enough of harmful or immature or downright silly philosophical ideas – is a fact that might well depress and even terrify us. And indeed, it would be quite true to say that almost all our wars are ideological wars: religious wars or ideological-religious persecutions.

But we must not be too pessimistic. Fortunately there are also good, humane, and wise philosophical ideas. And they, too, are powerful. There is, first of all, the idea of religious tolerance and of respecting opinions which differ from our own. And there are the philosophical ideas of justice and liberty. Countless men have sacrificed their lives for them. And if we mention ideological wars we must not forget such crusades of peace as the Nansen Aid of the International Red Cross in Geneva which saved more than a million citizens of the Soviet Union from death by famine in the years 1921 and 1922. And we must not forget that the idea of Peace on Earth is a philosophical as well as a religious idea, and that it was a philosopher, Immanuel Kant, who first formulated the idea of a World Federation of a League of Nations.

The idea of peace is a good example of our thesis of the political power of ideas. Obsessed as we are by the memory of two world wars and by the threat of a third, we all are inclined to overlook something important – the fact that since 1918 all Europe has acknowledged the idea of peace as fundamental. Even Mussolini and Hitler, whose ideology was openly aggressive, were forced, by the prevailing public opinion, to pose as friends of peace, and to blame others for the wars which they began. The fact that they had to make this concession to public opinion shows how strong was the will to make peace. One must not underrate the moral victory won in 1918 by the idea of peace. It has not, it is true, brought us peace. Yet it has created the *will* to make peace – a will which is a moral prerequisite for it.

This victory of the idea of peace can be regarded as a belated victory of Erasmus of Rotterdam, almost four hundred years after his death. To see clearly how badly Christian Europe stood in need of the teachings of the Christian humanist Erasmus, we should remember the attack made upon Erasmus by that great musician, poet, and fighter against the devil, Martin Luther.

Luther fought Erasmus because he saw that the idea of peace was linked with the idea of tolerance: 'If I did not see these upheavals [Luther speaks of war and bloodshed] I should say that the word of God was not in the world. But now, when I do see them, my heart rejoices …'. 'The wish to quieten these upheavals is nothing less than the wish to abandon the word of God and to suppress it', Luther wrote. And to Erasmus' appeal for peace and understanding he replied: 'Stop lamenting, stop trying to cure [the ills of the world]! This war is the war of our Lord God. He started it, He sustains it, and it will never cease until all the enemies of his word have become dung under our feet.'[12] We should remember here that Erasmus and his friends were not lacking in personal courage. Sir Thomas More and John Fisher, both friends of Erasmus and like himself champions of tolerance, died, not primarily as martyrs of Roman Catholicism, but rather, I believe, as martyrs of the idea of humanism, as opponents of barbarism, of arbitrary rule, and of violence. If today we look upon Christianity as a force for peace and tolerance, we testify to the spiritual victory of Erasmus.

III

All I have said so far was meant to suggest to you an attitude towards philosophy which might perhaps be formulated like this. Just as there are good and evil religions – religions that encourage the good or the evil in man – so there are good and evil philosophical ideas, and true and false philosophical theories. We must therefore neither revere nor revile religion as such, or philosophy as such. Rather, we must evaluate religious and philosophical ideas with critical and selective minds. The terrifying power of ideas burdens all of us with grave responsibilities. We must not accept or refuse them unthinkingly. We must *judge them critically.*

The attitude I have just formulated may appear to many as obvious. But it is not by any means generally accepted, or even generally understood. It is in origin, rather, a specifically European or Western attitude – *the attitude of critical rationalism.* It is the attitude of the critical and rational tradition of European philosophy.

Of course, critical thinkers have existed outside Europe. But nowhere else, to my knowledge, has there existed a critical or rationalist *tradition.* And from the critical or rationalist tradition in Europe there grew, eventually, European science.

But even before it gave rise to modern science, critical rationalism created European philosophy. Or more precisely, European philosophy is just as old as European critical rationalism. For both were founded by Thales and Anaximander of Miletus.

Naturally, uncritical and even anti-critical countercurrents – both rationalist and anti-rationalist – kept forming within European philosophy itself. And nowadays[13] the anti-rationalist philosophy of 'existentialism' is enjoying a great vogue.

Existentialism maintains, quite correctly, that in matters of real importance nothing can be proved, and that therefore one is always faced with the necessity of making decisions – *fundamental decisions*. But hardly anybody, not even the most uncritical and naïve rationalist, would contest the assertion that nothing of importance can be proved, and that all that can be proved consists, at most, of mathematical and logical truisms.

It is, therefore, perfectly correct to state that we have to make free decisions all the time – a fact that, for instance, Immanuel Kant, the critical rationalist and the last great philosopher of the Enlightenment, saw very clearly. But of course this statement tells us nothing about what our fundamental decision will turn out to be: whether we will decide for or against rationalism, whether, that is, we will decide with Erasmus and Socrates in favour of listening to rational arguments – making our further decisions depend on critical and careful consideration of such arguments, and on self-critical reflection – or whether we leap headlong into the magic circle of an irrationalist existentialism, or rather, into the magic whirlpool of anti-rationalist 'commitments'.[14]

However that may be, European philosophy made a fundamental decision, fundamental for itself and for Europe, when twenty-four centuries ago it decided in favour of critical rationalism and self-criticism. And indeed, without this self-critical tradition the current fashions of philosophical anti-rationalism could not possibly have arisen: it is just one of the traditions of critical rationalism that it never ceases to criticize itself.

IV

What I have said thus far bears heavily upon my main subject. Yet it is only an introduction. For the task of outlining the influence of European philosophy on the history of Europe in one short hour confronted me with some difficult and fairly fundamental decisions. I decided to limit myself to three closely connected problems. I want to discuss the little-known role which a highly immature philosophical theory has played in the rise of the three most distinctive and characteristic forces in European history. The three forces I have in mind are the following:

1 our industrial civilization;
2 our science, and its influence; and
3 our idea of individual freedom.

Thus industrialism, science, and the idea of freedom are my three chief subjects. It is fairly obvious that they are characteristically European subjects, provided we permit ourselves to treat American civilization as an offshoot of European civilization. How they link up with philosophy is perhaps less obvious.

It is my basic proposition that they are connected in an interesting way with a highly characteristic European theory of knowledge or epistemology: with that theory which Plato described in his famous simile of the cave in which he depicted the world of phenomena as a world of shadows – shadows cast by a real world hidden behind the world of phenomena. Admittedly Plato's belief in a world we can never learn to know could perhaps be called *'epistemological pessimism'*. And it has spread far beyond Europe. But Plato supplemented it, quite in the spirit of the old Ionian critical and rationalist tradition, with an unequalled *epistemological optimism*. And this epistemological optimism has remained part of our Western civilization. It is the optimistic theory that science, that is, real knowledge about the hidden real world, though certainly very difficult, is nevertheless attainable, at least for some of us. Man, according to Plato, can discover the reality hidden behind the world of phenomena, and he can discover it by the power of his own critical reason, without the aid of divine revelation.

This is the almost unbelievable optimism of Greek rationalism:[15] of the rationalism of the Renaissance – of European rationalism. Homer, though perhaps with a slight touch of irony, still appealed to the authority of the Muses. They are his sources, the divine fountainheads of his knowledge. Similarly the Jews and, in the Middle Ages, the Arabs and the Christians of Europe, appealed to the authority of divine revelation as the source of their knowledge. But the Ionian philosophers, beginning perhaps with Thales, *argued*. They appealed to *critical argument* and thus to reason: they thought of reason as capable of unveiling the secrets of a hidden reality. This is what I call 'epistemological optimism'. I believe that this optimistic attitude existed almost exclusively in Europe: in the two or three centuries of Greek rationalism and in the three or four centuries of its European and American renaissance.

Corresponding to my three main subjects, industrialism, science, and freedom, I can now formulate my three chief theses. Summed up in one sentence they read:

Europe's industrialism, its science, and its political idea of freedom, that is to say, every one of those characteristic and fundamental aspects of European civilization that I have listed, is a product of what I have called 'epistemological optimism'.[16]

I shall now try to substantiate this thesis for each of my three chief subjects.

V

When we seek to understand the distinctive character of European or Western civilization, one feature leaps to the mind. European civilization is an industrial civilization. It is based upon industrialization on the grand scale. It uses engines, sources of energy which are non-muscular. In this, European and American civilization differ fundamentally from all the other great civilizations which are or were mainly agrarian and whose industry depended on manual labour.

I think there is no other feature that distinguishes our civilization so clearly from all the others – except perhaps European science. Literature, art, religion, philosophy, and even the rudiments of natural science play their part in all other civilizations, for instance in those of India and China. But heavy industry on a grand scale seems to be unique as a form of production and indeed as a way of life. We find it only in Europe and in those parts of the world which have taken it over from Europe.

Like industrialism, the growth of science is a characteristic feature of Europe. And since they have developed almost simultaneously, the question arises whether industry is a product of the development of science or whether (as Marxism would have it) science is a product of industrialization.

I believe that neither of these interpretations is true, and that both science and industry are products of that philosophy which I have called 'epistemological optimism'.

It is a fact that ever since the Renaissance the development of industry and the development of science have been closely linked and have closely interacted. Each is indebted to the other. But if we ask how this interaction came about, my answer is this. It was bound to take place right from the outset, for it stemmed from a new philosophical or religious idea – a peculiar new variant of the Platonic idea that the philosophers, that is, those who know, should also be those who wield the power. The peculiar new variant of that theory is expressed in the dictum that *knowledge is power* – power over nature.

It is my thesis that both the industrial and the scientific development which took place since the Renaissance are realizations of this philosophical idea – *the idea of the mastery of man over nature*.

The idea of mastery over nature is, I suggest, the Renaissance version of epistemological optimism. We find it in the Neo-Platonist Leonardo, and in a somewhat claptrap form we find it in Bacon. Bacon was not, I believe, a great philosopher. But he was a visionary, and he is most important as the prophet of a new industrial and scientific society. He founded a new secularized religion and so became the creator of the industrial and scientific revolution.[17]

VI

Before going into details I would like to explain briefly my own opinion about this particular version of epistemological optimism.[18]

I am myself a rationalist, and an epistemological optimist. Yet I am no friend of that mighty rationalist religion of which Bacon is the founder. My objection to this religion is purely philosophical. And I should like to emphasize that it has nothing whatever to do with the present hangover – the intellectual anti-climax of the nuclear bomb[19] (or with other undesirable unintended consequences of the growth of scientific knowledge and technology). My objection to the religion of mastery over nature, to the idea that knowledge is power, is, quite simply, that knowledge is something far better than power. Bacon's formula 'knowledge is power' ('*nam et ipsa scientia potestas est*') was an attempt to advertise knowledge. It takes it for granted that power is always something good, and it promises that you will be repaid in terms of power if you make the unpleasant effort needed to attain knowledge. Yet I believe that Lord Acton was right when he said: 'Power tends to corrupt; and absolute power corrupts absolutely.'[20] Of course, I do not deny that power can be tamed, that it may sometimes be used for very good things – for instance, in the hands of a good physician. But I am afraid that even physicians not infrequently succumb to the temptation to make their patients feel their power.

Kant once commented strikingly on the saying that truthfulness, and honesty, is the best policy. This, he said, was open to doubt. But he added that he did not doubt that truthfulness was better than any policy.[21] My remark that knowledge is better than all power is merely a variant of this remark of Kant's. To the scientist only truth matters, not power. It is the politician who cares about power.

The idea of mastery over nature is in itself perhaps neutral. When it is a case of helping our fellow men, when it is a case of medical progress or of the fight against starvation and misery, then of course I welcome the power we owe to our knowledge about nature. But the idea of mastery over nature often contains, I fear, another element – the will to power as such, the will to dominate. And to the idea of domination I cannot take kindly. It is blasphemy, sacrilege, *hubris*. Men are not gods and they ought to know it. We shall never dominate nature. The mountaineer is to be pitied who sees in mountains nothing but adversaries he has to conquer – who does not know the feeling of gratitude, and the feeling of his own insignificance in the face of nature. Power always is temptation, even power or mastery over nature. What the Sherpa Tenzing felt on the peak of Chomo Lungma – that is, Mount Everest – was better: 'I am grateful, Chomo Lungma', he said.[22]

But let us return to Bacon. From a rational or critical point of view Bacon was not a great philosopher of science. His writings are sketchy and pretentious, contradictory, shallow, and immature. And his famous and influential theory of induction, so far as he developed it (for most of it remained a mere project, and has remained so ever since), bears no relation to the real procedure of science. (It was the great chemist Justus Liebig who pointed this out most forcefully.[23]) Bacon never understood the theoretical approach of Copernicus, or of Gilbert, or of his contemporaries Galileo and Kepler.[24] Nor did he understand the significance of mathematical ideas for science. Yet hardly any philosopher of modern times can compete with Bacon's influence. Even today many scientists still regard him as their spiritual father.

VII

This leads us to a question which I call the historical problem of Bacon: how can we explain the immense influence of this logically and rationally quite unimportant philosopher?

I have already briefly hinted at my solution of this problem. In spite of everything I have said, Bacon *is* the spiritual father of modern science. Not because of his philosophy of science and his theory of induction, but because he became the founder and prophet of a rationalist church – a kind of anti-church. This church was founded not on a rock but on the vision and the promise of a scientific and industrial society – a society based on man's mastery over nature. Bacon's promise is the promise of *self-liberation of mankind through knowledge*.[25]

In his Utopia, *The New Atlantis*, Bacon depicted such a society. The governing body of that society was a technocratic institute of research which he called 'Salomon's House'. It is interesting to note that Bacon's *New Atlantis* not only anticipated certain not too pleasant aspects of modern 'Big Science', but goes beyond them in its uninhibited dreams of the power, glory, and wealth that the great scientist may attain. Consider Bacon's description of the more than papal pomp of one of the 'Fathers of Salomon's House' – that is to say, one of the Directors of Research:[26]

The day being come, he made his entry. He was a man of middle stature and age, comely of person, and had an aspect as if he pitied men. He was clothed in a robe of fine black cloth, with wide sleeves and a cape. ... He had gloves that were ... set with stone; and shoes of peach-coloured velvet. ... He was carried in a rich chariot without wheels, litter-wise; with two horses at either end, richly trapped in blue velvet embroidered; and two footmen on each side in the like attire. The chariot was all of cedar, gilt, and adorned with crystal; save that the fore-end had pannels of sapphires, set in borders of gold, and the hinder-end the like of emeralds of the Peru colour. There was also a sun of gold, radiant, upon the top, in the midst; and on the top before, a small cherub of gold, with wings displayed. The chariot was covered with cloth of gold tissued upon blue. He had before him fifty attendants, young men all, in white sattin loose coats to the mid-leg; and stockings of white silk; and shoes of blue velvet; and hats of blue velvet; with fine plumes of divers colours, set round like hat-bands. Next before the chariot went two men, bare-headed, in linen garments down to the foot, girt, and shoes of blue velvet; who carried the one a crosier, the other a pastoral staff like a sheep-hook; neither of them of metal, but the crosier of balm-wood, the pastoral staff of cedar. Horsemen he had none, neither before nor behind his chariot: as it seemeth, to avoid all tumult and trouble. Behind his chariot went all the officers and principals of the Companies of the City. He sat alone, upon cushions of a kind of excellent plush, blue: and under his foot curious carpets of silk of divers colours, like the Persian, but far finer. He held up his bare hand as he went, as blessing the people, but in silence.

But a less questionable passage from Bacon's *Novum Organum* may here be of interest:[27]

> Also our hopes are raised by the fact that some of the experiments made hitherto are of such a kind that nobody before had an idea of such things; rather, they would have been contemptuously set aside as impossibilities.
>
> If prior to the discovery of firearms somebody would have described their effects, and would have said that an invention had been made by the means of which even the biggest walls and fortifications could be shaken and knocked down from a great distance, people might have, quite reasonably, deliberated about the various ways of utilizing the power of the existing machines and contraptions, and how one might strengthen them with more weights and more wheels, or increase the number of knocks and blows; but nobody would have dreamt of a fiery blast which suddenly and violently expands and blows up; one would, on the contrary, have discarded such a thing entirely, because nobody had ever seen an example of it.

Bacon then goes on to discuss in a similar vein the discovery of silk and of the mariner's compass – and he continues:

> Thus there is much hope that Nature still holds many an excellent and useful thing in her lap that has no resemblance or parallel to what has been discovered hitherto, but, on the contrary, lies far away from all the paths of imagination, and from all that has been found so far. No doubt it will come to light in the circuitous course of the centuries, just as has happened with earlier inventions; but with the aid of the method which is here treated of these things will be found far more surely and more quickly; and indeed, they might be accounted for and anticipated at once.

This passage from the *Novum Organum* is characteristic of Bacon's promise: Follow my new way, my new method, and you shall quickly attain knowledge and power. Indeed, Bacon believed that an Encyclopedia containing a description of all important phenomena of the Universe could be completed soon: he believed that, given two or three years, he could read through the whole book of Nature and bring the task of the new science to its completion.

There is no need to say that Bacon was mistaken – not only about the magnitude of the task, but about his new method. The method he was proposing had nothing whatever to do with that of the new science of Gilbert, of Galileo, or of Kepler, or with the later discoveries of Boyle and of Newton.

Yet Bacon's promise of a scientific future, splendid and close at hand, had an immense influence on both English science and the English industrial revolution – the industrial revolution which spread first to Europe and later to America and, indeed, all over the world, and which has truly changed the world into a Baconian Utopia.

As is well known, the *Royal Society* and later the *British Association for the Advancement of Science* (and still later the *American Association*) were deliberate attempts to give effect to the Baconian idea of cooperative and organized research. And it may be of interest here to quote a passage from the second charter of the Royal Society, of 1663, which is still in force. It says that the researches of the Members are to promote 'by the authority of experiments the sciences of natural things and of useful arts [that is, industrial technology], to the glory of God the Creator, and the advantage of the human race'.[28] The conclusion of this passage is taken almost literally from Bacon's *The Advancement of Learning*.[29]

Thus this pragmatic-technological attitude was combined from the outset with humanitarian aims: the increase of general welfare and the fight against want and poverty. The English and European industrial revolution was a philosophical and religious revolution, with Bacon as its prophet. It was inspired by the idea of accelerating, through knowledge and research, the hitherto far too slow advance of technology. It was the idea of a *material* self-liberation through knowledge.

VIII

But here one could raise an important objection. Was not the idea of applied knowledge, the idea that knowledge is power, already an influence in the Middle Ages? Was there not astrology that served the desire for power, and alchemy, the search for the Philosopher's Stone?

The objection is important and can help us to bring out more clearly the influence of epistemological optimism. For that peculiar optimism was lacking in the medieval alchemists and astrologers. They searched for a secret which, they believed, had once been known in antiquity but had later been forgotten. They looked for the key to wisdom in old parchments.

And yet, they may have been right in hunting for lost treasures of wisdom. What they were seeking so eagerly may well have been, unknown to themselves, the greatness of ancient Rome and the Augustan Peace, or perhaps the greatness and the boldness of the critical and rationalist philosophy of the pre-Socratic philosophers.

However this may be, Bacon (and the Renaissance) felt differently on this point. Admittedly Bacon was an alchemist and a 'magus' who believed in 'natural magic'. But he also believed – and this is decisive – that he himself had found a key to new wisdom. It is this new self-confidence that distinguishes Bacon's optimism – the confidence, completely unwarranted in his case, that he himself had the power to unveil the mysteries of nature without having to be initiated into the secret wisdom of the ancients. This power is independent of divine revelation, and independent of the disclosure of mysteries in the secret writings of ancient sages. Thus, Bacon's promise may be said to encourage enterprise and self-confidence. It encourages men to rely on themselves in the search for knowledge, and so to become independent of divine revelation and of ancient traditions.

IX

Bacon himself (and with him many other Renaissance sages) belonged to two worlds: he belonged to the old world of mysticism and word magic combined with the authoritarian faith in some lost secret, the (Neo-Platonic[30]) *Wisdom of the Ancients*.[31] At the same time he belonged to the new world of an anti-authoritarian confidence in our own power of adding to our wisdom and thereby of further increasing our power. This made Bacon's prophetic message apt to grow into a new religion, and ultimately into the new message of the Enlightenment. This new message of the European Enlightenment might perhaps be summed up in the somewhat equivocal formula: *God helps those who help themselves* – a formula which has sometimes been taken quite seriously as a statement of the responsibility imposed by God on us, and sometimes as a manifesto of the self-emancipation and the self-reliance of a secular, fatherless society.[32]

Christianity, perhaps more than any other religion, had always taught its believers to look forward to a life to come, to sacrifice the present for the sake of the future. Thus it laid the foundation of an attitude towards life that might be called the European 'futurity neurosis'. It is a way of living at all times more in the future than in the present, of being obsessed with plans for the future, schemes for the future, investments in a better life to come. My conjecture is that epistemological optimism with its peculiar idea of self-reliance – that God helps those who help themselves – secularized Christianity, and turned its futurity neurosis into the idea of self-liberation through the acquisition of new knowledge and through partici-pation in the new knowledge to come – the new growth of knowledge – and at the same time into the related but subtly different idea of self-liberation through the acquisition of new power, and of new wealth.

Thus we may say that the Baconian Utopia, like most Utopias, was an attempt to bring heaven down to earth. And so far as it promised an increase of power and of wealth through self-help and self-liberation through new knowledge, it is perhaps the one Utopia that has (so far) kept its promise. Indeed it has kept it to an almost unbelievable extent.[33]

X

I may perhaps now remind you of my programme, which was to outline the decisive part played by philosophical ideas, and more precisely by epistemological optimism, in the development of three characteristic forces of European history:

1 our industrial civilization;
2 our science, and its influence; and
3 our idea of individual freedom.

I will now leave the first of those three points – not because I have exhausted it (it is a subject I could not exhaust in one lecture), but simply because I have to pass on to my second point, the evolution of modern science.

As I have pointed out before, the evolution of science and of industry and technology have interacted, and have enriched each other. I now only want to stress that this interaction shows a significant asymmetry. While modern industrial development has become unthinkable without modern science, the opposite is not the case: science is largely autonomous. No doubt the needs of industry have been a stimulus to its development, and any stimulus is welcome and useful. But what the scientist wants more than anything else is to know. And though he is grateful to anyone who gives him interesting problems to tackle, and the means to tackle them, what he wants is knowledge, and to add to our knowledge.

XI

The science of the Renaissance may be regarded as the direct continuation of the Greek cosmology of the Ionians and the Pythagoreans, the Platonists and the Aristotelians, the Atomists and the Geometers. The method of Galileo and of Kepler is the critical, the rational, the hypothetical method of these forerunners of theirs.[34] Hypotheses are invented and criticized. Under the influence of criticism they are modified. When the modifications become unsatisfactory, they are discarded, and new hypotheses are advanced. One typical example is the Ptolemaic geocentric cosmology with its modifications and auxiliary hypotheses, the epicycles. When they became too cumbrous, Copernicus rediscovered the heliocentric cosmology of Aristarchus. The heliocentric hypothesis, too, led to grave difficulties. But they were triumphantly resolved by Kepler and Newton. Thus the method of science consists of boldly advancing tentative hypotheses and of submitting them to critical tests. Since Einstein we know that it can never lead to certainty. For whether Newton or Einstein is right, we have learned from Einstein at least one thing: that Newton's theory, too, is only a hypothesis, a conjecture, and perhaps a false one, in spite of its incredible success in predicting with the greatest precision almost all astronomical phenomena within our solar system, and even beyond.

Thus we have learned from Einstein that science offers us only hypotheses or conjectures rather than certain knowledge. But the modest programme of searching for hypotheses would probably not have inspired scientists: it might have never started the enterprise of science. What people hoped for, and sought, was knowledge – certain indubitable knowledge. Yet while searching for certain knowledge, scientists stumbled, as it were, upon the hypothetical, the conjectural, the critical method. For knowledge, whether certain or not, had to stand up to criticism. If it failed to do so, it had to be discarded. And so it came about that scientists got used to trying out new conjectures, and to using their imagination to the utmost while submitting themselves to the discipline of rational criticism.

Although nowadays we have given up the idea of absolutely certain knowledge, we have not by any means given up the idea of the search for truth. On the contrary, when we say that our knowledge is not certain, we only mean that we can never be sure whether our conjectures are *true*. When we find that a hypothesis is not true, or at least that it does not appear to be a better approximation to the truth than its competitors, then we may discard it. Hypotheses are never verifiable, but they can be falsified. They can be criticized, and tested.

It is the search for true theories that inspires this critical method. Without the regulative idea of truth, criticism would be pointless.

The experimental methods of Gilbert, Galileo, Torricelli, and Boyle are methods for testing theories: if a theory fails to satisfy an experiment, it is falsified and has to be modified or supplanted by another one – by a theory, that is, which lends itself better, or at least equally well, to testing.

So much for the method of science. It is critical, argumentative, and *almost* sceptical.

XII

But the great masters of this method were not aware of the fact that this was their method. They believed in the possibility of reaching absolute certainty in knowledge. A radical version of epistemological optimism inspired them (as it inspired Bacon). It led them from success to success. Nevertheless it was uncritical, and logically untenable.

We can describe this radical and uncritical epistemological optimism of the Renaissance as the belief that *truth is manifest*. Truth may be hard to find. But once the truth stands revealed before us, it is impossible for us not to recognize it as truth. We cannot possibly mistake it. Thus nature is an open book. Or, as Descartes put it, God does not deceive us.

This theory is closely related to Plato's theory of anamnesis – to the theory that before our birth we knew the hidden reality, and that we recognize it again once we happen to catch sight of it, or perhaps even its faint shadow.

The idea that truth is manifest is a philosophical idea (or perhaps even a religious idea) of the greatest historical importance. It is an optimistic idea, a beautiful and hopeful dream, a truly sublime idea. And I willingly admit that there may be a grain of truth in it. But certainly not more than a grain. For the idea is mistaken. Again and again, even with quite simple things, we hold the truth in our hands and do not recognize it. And still more often we are convinced of having recognized the manifest truth while in fact we are entangled in errors.

The radical epistemological optimists – Plato, Bacon, Descartes, and others – were of course aware of the fact that we sometimes mistake error for truth. And in order to save the doctrine of manifest truth they were forced to explain the occurrence of error.

Plato's theory of error was that our birth is a kind of epistemological fall from grace: when we are born we forget the best part of our knowledge, which is our direct acquaintance with truth. Similarly, Bacon (and Descartes) declared that error is to be explained by our personal shortcomings. We err because we stubbornly cling to our prejudices instead of opening our physical or mental eyes to the manifest truth. We are epistemological sinners: hardhearted sinners who refuse to perceive the truth even when it is manifest before our eyes. Bacon's method, therefore, consists in cleansing our minds of prejudices. It is the unbiased mind, the pure mind, the mind cleansed of prejudice, that cannot fail to recognize the truth.

With this theory I have reached a final formulation of radical epistemological optimism. The theory is of great importance. It became the cornerstone of modern science. It made the scientist a priest of truth, and the worship of truth a sort of divine service.

I believe that this respect for truth is indeed one of the most important and most precious traits of European civilization, and that it is rooted nowhere more firmly and deeply than in science. It is a priceless treasure we find in the treasure-house of science, a treasure which, I believe, far surpasses her technological utility.

Yet Bacon's theory of error is, in spite of its desirable consequences, untenable. Small wonder, therefore, that it has also led to undesirable consequences. I shall discuss some of these consequences in connection with my third and last point to which I am now coming – my analysis of the importance of epistemological optimism for the development of freedom, of European liberalism.

XIII

In discussing my second point I have tried to show in what way epistemological optimism is responsible for the development of modern science. At the same time I have tried to discuss epistemological optimism and to evaluate and criticize this peculiar philosophy.

All this we have to keep in mind when we now turn to glance at the development of modern liberalism. And since I am going to say something critical about it, I want first of all to state quite clearly and unmistakably that my sympathies are all for it. Indeed, while I am well aware of its many imperfections, I do think – with E.M. Forster and Pablo Casals – that democracy is the best and noblest form of social life that has so far arisen in the history of mankind. I am not a prophet, and I cannot deny the possibility that one day it will be destroyed. But whether or not it will in fact survive, we should work for its survival.

Now I think that the mainspring that keeps democratic societies going is the peculiar philosophy that I have just sketched: the belief in the sanctity of truth, together with the over-optimistic belief that truth is manifest, even though it can be temporarily obscured by prejudices.

This peculiar philosophy is, of course, much older than Bacon. It played a large part in almost all religious wars – each side regarding the other as benighted, as refusing to see the obvious truth, and perhaps even as possessed by the devil.

XIV

Epistemological over-optimism has two very different philosophies opposed to it: a pessimism which despairs of the possibility of knowledge, and a critical optimism which realizes that it is human to err and that fanaticism is usually the attempt to shout down the voices of one's own doubts. Up to the twentieth century critical optimists were rare. Socrates, Erasmus, John Locke, Immanuel Kant, and John Stuart Mill were among the greatest of them.

The development of liberalism from the Reformation to our own time took place almost entirely under the sway of an uncritical, epistemological over-optimism: the theory of manifest truth. This theory led liberalism along two roads. The first led straight from the Reformation to the demand for freedom of religious worship. The second led through some disappointments in the theory of manifest truth to the theory that there exists a conspiracy against truth. For, it was argued, if so many do not see the manifest truth – that truth which is so clearly visible – it must be on account of false prejudices cunningly and systematically implanted into young impressionable minds so as to blind them to truth. The conspirators against truth were, of course, the priests of the competing churches: in the minds of the Protestants the Catholic Church, and *vice versa*.

Though based upon the mistaken doctrine of manifest truth, this second road led, nevertheless, to the valid and invaluable demand for freedom of thought, and to the demand for a universal and secular primary education – on the ground that those who are freed from the darkness of illiteracy and religious tutelage cannot fail to see the manifest truth.

And it led finally to the demand for universal suffrage. For if truth is manifest, then people cannot err. And since they can recognize the truth, they can also recognize what is good and just.

I believe that this development was good and just – despite the epistemological over-optimism, which is the main weakness of its theoretical basis. But it was the weakness of this theoretical basis that led to the terrible religious wars of the sixteenth and seventeenth centuries, and to the horrors of violent revolutions and civil wars. Here in the West, all this has finally led us to the Socratic insight that to err is human. We are not fanatics any longer. Most of us are only too willing to recognize our own shortcomings and mistakes. This insight, belatedly as it came to us, is a blessing. Like all blessings, though, it is a mixed blessing, for it tends to undermine our confidence in our way of life, especially those of us who have learned this lesson well.

I have come to the end of my historical sketch, and in conclusion I wish to add only one further remark: the Socratic-Erasmic insight that we may be in error should certainly prevent us from waging a war of aggression. But our consciousness of our shortcomings and mistakes must not deter us from fighting in defence of freedom.

NOTES

1 Plato, *Republic*, 473 c–e. In my *Open Society*, chapter 8, pp. 151f., I translated the passage as follows:

Unless, in their cities, philosophers are vested with the might of kings, or those now called kings and oligarchs become genuine and fully qualified philosophers; and unless these two, political might and philosophy, are fused (while the many who nowadays follow their natural inclination for one of these two are suppressed by force), unless this happens, my dear Glaucon, there can be no rest; and the evil will not cease to be rampant in the cities – nor, I believe, in the race of men.

2 See T. Masaryk, *The New Europe (The Slav Standpoint)*, printed by Eyre and Spottiswoode, London, 1918, for private circulation, p. 68.
3 It might, perhaps, be mentioned that Masaryk's nationalism was moderate and humane: 'I have never been a Chauvinist; I have not even been a nationalist ...' (Masaryk, p. 45). Nevertheless, he also said: '... we advocate the principle of nationality ...' (Masaryk, p. 52) and demanded the dismemberment of Austria-Hungary into national states. (See also note 53 to chapter 12 of my *Open Society*.) For a most interesting though different appraisal of Masaryk's views, see A. van den Beld, *Humanity: The Political and Social Philosophy of Thomas G. Masaryk*, Mouton, The Hague, 1975.
4 Iceland may be an exception. Cp. my *Conjectures and Refutations*, p. 368.
5 A.J.P. Taylor claims that 'Czechoslovakia contained seven [nationalities]' – that is, 'Czechs, Slovaks, Germans, Magyars, Little Russians, Poles, Jews'. See *The Hapsburg Monarchy*, (1948), Peregrine Books edition, 1964, p. 274.
6 Heinrich Heine, *Zur Geschichte der Religion und Philosophie in Deutschland*, 1833–4, book III (see p. 150 of Wolfgang Harich's edition, Insel Verlag, Frankfurt, 1966). The passage was quoted in my *Open Society*, volume II, p. 109.
7 Hayek has long stressed this point which, he says, has 'long formed a fundamental part of the liberal creed'. See his *Constitution of Liberty*, Routledge & Kegan Paul, London, 1960, pp. 112f., the quotation in his text from J.S. Mill, and in his footnote 14 on p. 445 from Lord Keynes.
8 Marx does so in the 'Author's Preface to the Second German Edition' of *Capital*, dated London, 24 January 1873; see p. 871 of the Everyman edition of *Capital*, volume I; p. 27 of the Lawrence & Wishart edition, volume I.
9 This brief account is based on the analysis of Marx's theory that I give in my *Open Society*, chapters 13–21. See especially chapter 15, pp. 108f., footnote 13 on p. 326, and the references given there to Marx's 'Preface' to his *A Contribution to the Critique of Political Economy* and his *The Poverty of Philosophy*.
10 Cp. my *Open Society*, volume II, pp. 83 and 108.
11 See John Stuart Mill, *Autobiography*, chapter 4, first edition, 1873, p. 105, Houghton Mifflin/Oxford University Press edition, edited by J. Stillinger, 1969, 1971, p. 64. It seems that Mill believed at the time that the only means of realizing his programme was by the voluntary adoption of birth control by the 'labouring classes'. (The quotation in the text continues: 'through a voluntary restriction of the increase of

their numbers'.) There is no reason to think that he gave up his support for birth control. But in his *Autobiography*, chapter 7 (1st edition, p. 231; Stillinger edition, p. 138) he indicates (possibly under the influence of his wife: note the sudden 'we' instead of 'I' on the page referred to) that a necessary additional means was a changed attitude towards private property, and the adoption of a form of socialism.

12 The quotations are from Martin Luther's *De servo arbitrio (The Bondage of the Will)*, 1525, a book written in reply to Erasmus' *De libero arbitrio (A Diatribe or Discourse Concerning Free Choice)*, 1524. The translations are my own. See *De servo arbitrio* in *D. Martin Luther's Werke, Kritische Gesamtausgabe* (Weimarer Ausgabe), 18. Band, Hermann Böhlans Nachfolger, Weimar, 1908, p. 626. Cp. *Luther's Works, volume 33, Career of the Reformer III*, Fortress Press, Philadelphia, 1972, pp. 52–3; or Martin Luther, *Ausgewählte Werke*, edited by H.H. Borcherdt and G. Merz, Ergänzungsreihe, erster Band, 1954, p. 35.

13 This passage was written in 1959.

14 Decisions ('making up our minds') are unavoidable, even in science. What scientists do all the time is to decide, in the light of argument. But we should distinguish between critical and tentative decisions and dogmatic decisions or commitments. It is the latter type of decision which has given rise to 'decisionism'.

15 It will be clear from the context that I am using the term 'rationalism' in its wide sense, and not in the narrow sense in which it is used in opposition to 'empiricism'.

16 For a discussion of epistemological optimism, see 'On the Sources of Knowledge and of Ignorance' in my *Conjectures and Refutations*, pp. 5ff.

17 Only quite recently, and many years after I first arrived at my not too favourable opinion of Bacon's philosophy as well as at the view that he was the prophet of the industrial revolution, I came across the admirable and highly original book by Benjamin Farrington, *Francis Bacon, Philosopher of Industrial Science*, (Schuman, New York, 1949; Collier, New York, 1961; Lawrence & Wishart, London, 1951). Though Farrington treats Bacon from a philosophical point of view very different from my own, our results concerning Bacon's influence on the industrial revolution are strikingly similar. Indeed, Farrington quotes (on p. 136 of the 1961 American edition) the passage from Marx's *Capital* which I have now adopted as a motto for this chapter. In that passage Marx says: '... Francis Bacon *looked forward* to an alteration in the form of production and to the effective control of nature by man, as a result of a *change in the ways of thinking*' (italics mine). I certainly agree with what Marx says here, though my interpretation hardly fits Marx's own view of the relation between 'the mode of production of material life' and 'the general character of the social, political and intellectual processes of life'. For Marx says in his 'Preface' to *A Contribution to the Critique of Political Economy* (cp. Lawrence & Wishart edition, 1971, pp. 20–1; Karl Marx, *Selected Writings in Sociology and Social Philosophy*, edited by T.B. Bottomore and

M. Rubel, Penguin edition, 1963, p. 67), 'The mode of production of material life determines the general character of the social, political and intellectual processes of life. It is not the consciousness of men that determines their existence but, on the contrary their social existence determines their consciousness.' This can hardly be squared with interpreting Bacon as one of those who helped to bring about the industrial revolution 'as a result of a change in the ways of thinking'.

18 Cp. my 'Science: Problems, Aims, Responsibilities', chapter 4 of the present volume.

19 In the original lecture I mentioned at this point that my critical attitude towards Bacon predated the creation of nuclear weapons (I criticized Bacon in 1934), and that I remained a great admirer of Albert Einstein and Niels Bohr even though their theories fathered the atom bomb.

20 Lord Acton, Letter to Mandell Creighton, 5 April 1887. Cp. Lord Acton, *Essays on Freedom and Power*, edited by Gertrude Himmelfarb, Meridian Books, Thames & Hudson, London, 1956, p. 335.

21 I referred to this in my *Open Society*, volume I, p. 139. See Immanuel Kant, *On Eternal Peace*, Appendix, in *Kants gesammelte Schriften*, edited by *königlich preussischen Akademie der Wissenschaften*, VIII, Gruyter, Berlin and Leipzig, 1923, p. 370. Cp. *Kant's Political Writings*, edited by H. Reiss, Cambridge University Press, Cambridge, 1971, p. 116: 'It is true, alas, that the saying *"Honesty is the best policy"* embodies a theory which is frequently contradicted by practice. Yet the equally theoretical proposition *"Honesty is better than any policy"* infinitely transcends all objections, and is indeed an indispensable condition of any policy whatever.'

22 Tenzing Norgay, *Man of Everest* (as told to James Ramsey Ullman), George Harrap & Co. Ltd, London, 1955, p. 271.

23 See Justus von Liebig, *Ueber Francis Bacon von Verulam und die Methode der Naturforschung*, Munich, 1863; English translation, 'Lord Bacon as Natural Philosopher' I and II, *Macmillan's Magazine, 8*, 1863, pp. 237–49, 257–67.

24 Kepler is not mentioned at all by Bacon. See Ellis' 'Preface' to the *Descriptio Globi Intellectualis*, in *The Works of Francis Bacon*, edited by James Spedding, Robert Leslie Ellis, and Douglas Denon Heath, Longman & Co., London, 1862–75, volume III, pp. 723–6; on Copernicus, see III, p. 229 and V, p. 517 (also IV, p. 373); on Galileo, see II, p. 596 (Bacon on Galileo's theory of the tides) and, for example, V, pp. 541–2; on Gilbert, see III, pp. 292–3 and V, p. 202 (also V, pp. 454, 493, 515 and 537).

25 Cp. my 'Emancipation Through Knowledge' in my *In Search of a Better World*.

26 Francis Bacon, *New Atlantis*, in *The Works of Francis Bacon*, volume III, pp. 114f.

27 The quotations are from *Novum Organum*, 109th Aphorism. Cp. *The Works of Francis Bacon*, volume I, pp. 207f. The translation is my own. Cp. James Spedding's translation in *The Works of Francis Bacon*, volume IV, pp. 99f.

28 Cp., for example, Sir Henry Lyons, *The Royal Society 1660–1940*,

Cambridge University Press, Cambridge, 1944, Appendix I: 'Second Charter: 22 April 1663'. The quotation in my text is from p. 329 of this Appendix.

29 See *The Works of Francis Bacon*, volume III, p. 294: '... for the glory of the Creator and the relief of man's estate'. Cp. also Rawley's introduction 'To the Reader' to *New Atlantis* (published in 1627): '... a college instituted for the interpreting of nature and the producing of great and marvellous works for the benefit of men ...' (*The Works of Francis Bacon*, volume III, p. 127.)

30 I should perhaps have said, rather, Hermeticism. See, notably, P. Rossi, *Francis Bacon: From Magic to Science*, Routledge & Kegan Paul, London, 1968 (first published in Italian in 1957); Francis Yates, *Giordano Bruno and the Hermetic Tradition*, Routledge & Kegan Paul, London, 1964; Francis Yates, 'The Hermetic Tradition in Renaissance Science', in *Art, Science and History in the Renaissance*, edited by C.S. Singleton, Baltimore, 1967. For a recent discussion see D.K. Probst, *Francis Bacon and the Transformation of the Hermetic Tradition into the Rationalist Church*, D.Sc. thesis, Université Libre de Bruxelles, Faculté des Sciences, Service de Chimie Physique II, 1972.

31 Cp. *De Sapientia Veterum*, in *The Works of Francis Bacon*, volume VI, pp. 619–86 (translation, pp. 689–764) and *De Principiis Atque Originibus Secundum Fabulas Cupidinis et Coele* in *The Works of Francis Bacon*, volume III, pp. 79–118 (translation, volume V, pp. 461–500), which interpreted a number of classical myths as cosmological allegories.

32 This term is a reference to the idea that our western societies do not, by their structure, satisfy a need for a father-figure. I discussed these problems, briefly, in my (unpublished) William James Lectures delivered in Harvard in 1950. (Cp. my *Conjectures and Refutations*, p. 375.)

33 I offer this view as an alternative historical conjecture to the theories of Max Weber and R.H. Tawney about the relation between 'Religion' and 'the Rise of Capitalism' (see Max Weber, *The Protestant Ethic and the Spirit of Capitalism*, G. Allen & Unwin, London, 1930; and R.H. Tawney, *Religion and the Rise of Capitalism*, Holland Memorial Lectures, 1922, first published 1926). The time at my disposal did not allow me to elaborate my conjecture – and even less to compare it critically with its competitors.

34 Lest I be misunderstood, my comments pertain not so much to scientists as individuals as to the scientific tradition – the friendly-hostile cooperation of scientists – which itself emerged from the very developments we are discussing. Cp. my *Objective Knowledge*, chapter 4, section 9.

EDITOR'S AFTERWORD

The Myth of the Framework was conceived in the mid-1970s. Several versions of eleven different essays were brought together, a working copy was made, typescripts were sent to reviewers, criticisms were solicited and received, revisions made, permissions requested – and then, the mounting pressure of other commitments forced Sir Karl to put the manuscript in his drawer. It sat there in typescript until 1986, when the Hoover Institution on War, Revolution and Peace acquired Popper's papers and created The Karl Popper Archives at their library in Stanford University.

The volume published here differs somewhat from what was planned in the mid-1970s. The greatest difference is the omission of 'Emancipation Through Knowledge' and 'The Logic of the Social Sciences', both of which have only recently appeared in Popper's *In Search of A Better World*. A third essay, 'The Frankfurt School', now forms an addendum to 'Reason or Revolution?', and 'Philosophy and Physics', not originally included, has now been added.

I have learned a great deal in editing this book from the work that Sir Karl has devoted to the essays that are published here, and, in particular, from the revisions that he has made to them. There are typically several versions of each of these essays in the archives, and typically several copies of each version. Each of these copies, moreover, typically has its own set of comments and corrections. Since few of them are dated, I at first found it difficult to determine which was intended as the latest version. But as time went on, it became clear that Sir Karl's revisions were always made in an effort to simplify his expression, so as to make his ideas more accessible to his readers. Sir Karl encouraged me to continue in the same vein, and to simplify his expression whenever it was possible to do so without altering his thought. I have not indicated the revisions that I have made in the body of the text, so as not to distract the reader's

210

attention. And since Sir Karl has approved them as his own, I see no reason to indicate them here.

I want to thank Sir Karl for trusting me with his work, and for the many stimulating conversations we have had concerning it.

The Myth of the Framework is the first volume to be published from The Karl Popper Archives. It is intended as the first of many. I want to acknowledge with gratitude the work that W.W. Bartley, III and Stanford University's Hoover Institution on War, Revolution and Peace have devoted to the creation of these archives.

In March of 1992 the Ianus Foundation began financial support of my work in the Popper Archives. Since then, it has supplied me with a microfilm copy of the archives and with the equipment necessary to use it. More important, it has also made possible an extensive transatlantic telephone tutorial with Sir Karl, as well as visits with him in the final stages of the preparation of this book. I want to thank the Scientific Director of the Ianus Foundation, Werner Baumgartner, for his vision of a Popper tree, and, most of all, for his friendship. And I want to thank the President of the Ianus Foundation, Jim Baer, who made sure that I had the best possible equipment for the job. I also want to thank Elisabeth Erdman-Visser (who first suggested to Sir Karl that I edit his work), Ursula Lindner and Melitta Mew (each of whom provided necessary moral support), Yvonne Damian (for her unflagging help in checking references), Richard Stoneman (Senior Editor at Routledge), Sue Bilton (who guided the book through its publication with Routledge), and Victoria Peters (who helped her to do it). In January of 1994, the Soros Foundation and the Central European University assumed the financial support of my project. I want to thank George Soros, both for his interest in my work, and for his commitment to Sir Karl's vision of the world as an open society. As I mentioned earlier, this volume was first planned for publication in the mid-1970s. Jeremy Shearmur worked on it then, and brought it a good way along the path toward publication. I owe a great debt of gratitude to him, since his work has undoubtedly made mine easier. Thanks are also due to Larry Briskman, Bryan Magee, and David Miller – each of whom made helpful comments on that original version of the text. Finally, I want to thank Kira Victorova, my wife and colleague, who helped in the proofreading, and in the preparation of the indexes of this book, and who has made my life while working on it a joy.

M.A. Notturno

Chicago, 1994

BIBLIOGRAPHY

The following is a list of Karl Popper's books in the English language

The Open Society and Its Enemies, Volume I: *The Spell of Plato*, Routledge & Kegan Paul, London, 1945, revised 1952, 1957, 1962, 1966.

The Open Society and Its Enemies, Volume II: *The High Tide of Prophecy: Hegel, Marx, and the Aftermath*, Routledge & Kegan Paul, London, 1945, revised 1952, 1957, 1962, 1966.

The Poverty of Historicism, Routledge & Kegan Paul, London, 1957.

The Logic of Scientific Discovery, Hutchinson, London, 1959; reprinted by Routledge, London, 1992. Translation of *Logik der Forschung*, Julius Springer, Vienna, 1934; revised edition J.C.B. Mohr, Tübingen, 1989.

Conjectures and Refutations: The Growth of Scientific Knowledge, Routledge & Kegan Paul, London, 1963, revised 1965, 1969, 1972, 1989.

Objective Knowledge: An Evolutionary Approach, Clarendon Press, Oxford, 1972, revised 1983.

Unended Quest: An Intellectual Autobiography, Open Court, La Salle, Illinois, 1982; revised edition published by Routledge, London, 1992. First published as *Autobiography of Karl Popper* in The Library of Living Philosophers, ed. Paul Arthur Schilpp, Open Court, La Salle, Illinois, 1974.

The Self and its Brain: An Argument for Interactionism, with John C. Eccles, Springer Verlag, Berlin, Heidelberg, and London, 1977; Routledge & Kegan Paul, London, 1983.

The Open Universe: An Argument for Indeterminism, Volume II of the *Postscript to The Logic of Scientific Discovery*, ed. W.W. Bartley, III, Hutchinson, London, 1982; paperback published by Routledge, London, 1988.

Quantum Theory and the Schism in Physics, Volume II of the *Postscript to The Logic of Scientific Discovery*, ed. W.W. Bartley, III, Rowman & Littlefield, Totowa, New Jersey, 1983; Unwin Hyman, London, 1982; reprinted by Routledge, London, 1992.

A Pocket Popper, ed. David Miller, Fontana, London, 1983; republished as *Popper Selections*, Princeton University Press, Princeton, New Jersey, 1985.

A World of Propensities, Thoemmes, Bristol, 1990.

BIBLIOGRAPHY

In Search of a Better World: Lectures and Essays from Thirty Years,
Routledge, London, 1992. Translation of *Auf der Suche nach einer
besseren Welt,* Piper, Munich, 1984, 1988.

The Myth of the Framework, ed. M.A. Notturno, Routledge, London,
1994.

Knowledge and the Body–Mind Problem, ed. M.A. Notturno, Routledge,
London, 1994.

NAME INDEX

SUBJECT INDEX

civilization/s 29, 38, 191–3, 200,
203; industrial 191, 193, 200;
result of culture clash 38;
western 29, 192–3, 203
clarity 70–1, 79
class struggle 53, 87
commitment: as irrationality 180
competition 7, 16, 27, 37, 54–5, 58,
60, 69, 86, 89–93, 102, 105, 109,
147, 160–1, 170, 182; among
theories 7, 16, 27, 37, 54, 55, 58,
60, 86, 89–92, 102, 105, 147, 160,
161, 170, 175, 178; basis of
objectivity 69, 93 (*see also*
discussion/s, critical)
confrontation/s: of different
cultures 38–40, 43, 51–3, 61,
170–1
conjecture/s *see* theories
conventions 37, 39–40, 45–6, 59,
61–2; and arbitrariness 45, 62;
and laws 45–6; and nature 37,
39–40, 61; truth not subject
to 39
conversion 51, 57
corroboration 181
cosmology/ies 20, 40–1, 43, 201; as
the basis of science 43; *see also*
astronomy
creativity 6, 9, 52, 71
critical: argument/s xii, xiii, 93–4,
142, 180, 191–2; attitude 51, 68,
70–1, 93–4, 103; discussion xiii,
8, 13, 34–7, 40–1, 43–4, 46, 48,
51, 54–6, 58–60, 69–70, 92–4, 98,
101, 104–5, 132, 140, 148,
158–63, 181; method/s 7, 30,
40–3, 58, 179, 201–2; optimism
204; rationalism xii, 190–2, 194;
tradition 38, 42, 61, 70, 72, 190–2
criticism xii, 7–8, 15–16, 29, 34,
40–3, 51–2, 54, 58–61, 68–70, 74,
80, 83, 86, 90, 92, 94, 96, 98, 103,
111, 117–18, 135–6, 140–2, 144,
147, 159–62, 180, 190–1, 201–2;
invention of 40–3; mistaken
method vs correct method 60;
mutual 34, 94; rational xii, 68–70,
147, 202; self- 15–16, 191;

transcendence through 51–2,
58–61
culture/s 37–40, 43, 51–3, 61, 170–1;
clash 38–40, 43, 51–3, 61, 170–1;
gulf between different 37–8; and
nationalism 186; value of culture
clash 51

Darwinism 7–9, 17, 19, 25, 29, 68,
90; social 17, 29
decision/s 191, 207; critical vs
tentative 207; necessity of
making 191, 207
decoding 24, 26
definition/s 59, 91
demarcation 29, 83–4, 88
democracy 40, 110, 127, 187, 204
determinism 11, 83, 108, 131, 181;
historical 131; macro-physical
11; scientific 83; theological 83
dialectic 70, 79, 136
dictatorship 185, 188; of the
proletariat 188
disagreement/s 34–7; desirability of
37; growth of knowledge
depends upon x, 34–6
discovery/ies 2–3, 5–7, 9, 12–13, 86,
92, 105–6, 128, 151; accidental vs
systematic 85, 106, 151
discussion/s 8, 13, 33–8, 40–1, 43–4,
46, 48, 51, 54–60, 69–70, 92–4,
98, 101, 104–6, 110, 117, 122,
132, 140, 148, 158–63, 181;
among scientists 55, 106;
common ground of 33; critical
(rational) xiii, 8, 13, 34–7, 40–1,
43–4, 46, 48, 51, 56, 58–60,
69–70, 92–4, 98, 101, 104–5, 132,
140, 148, 158–63, 181; critical
discussion is always possible 54;
ethical 122; fruitful 34–7, 54,
56–7; inconclusive 38; method of
43, 60; pessimism concerning
33–6, 47, 55
divine revelation 192, 199
dogmatism 14–17, 57, 59–60, 82–6,
94
dualism of matter and field 117
duty/ies xiii, 123, 126–7; of